MAGNESIUM ALLOYS - SELECTED ISSUE

Edited by **Tomasz Tański, Wojciech Borek and Mariusz Król**

Magnesium Alloys - Selected Issue
http://dx.doi.org/10.5772/intechopen.74194
Edited by Tomasz Tański, Wojciech Borek and Mariusz Król

Contributors

Rong-Chang Zeng, Zheng-Zheng Yin, Xiao-Bo Chen, Dao-Kui Xu, Andrzej Kielbus, Sebastian Feliu Batlle, Santiago Fajardo, Federico R García-Galván, Violeta Barranco, Juan Carlos Galván, Zuzanka Trojanova, Zdeněk Drozd, Pavel Lukac, Isao Nakatsugawa, Yasumasa Chino, Hideki Nakano, Chung-Hsing Chao, Tomasz Arkadiusz Tański

Notice

First published in London, United Kingdom, 2018 by IntechOpen
IntechOpen is the global imprint of INTECHOPEN LIMITED, registered in England and Wales, registration number: 11086078, The Shard, 25th floor, 32 London Bridge Street
London, SE19SG – United Kingdom
Printed in Croatia

British Library Cataloguing-in-Publication Data
A catalogue record for this book is available from the British Library

Additional hard copies can be obtained from orders@intechopen.com

Magnesium Alloys - Selected Issue, Edited by Tomasz Tański, Wojciech Borek and Mariusz Król
p. cm.
Print ISBN 978-1-78984-375-0
Online ISBN 978-1-78984-376-7

For EU product safety concerns:
IN TECH d.o.o., Prolaz Marije Krucifikse Kozulić 3, 51000 Rijeka, Croatia,
info@intechopen.com or visit our website at intechopen.com.

Meet the editors

Prof. Tomasz Tański is the Head of the Institute of Engineering Materials and Biomaterials, Silesian University of Technology, Poland. He is a specialist in materials such as non-ferrous alloys, composite and nanostructured materials, manufacturing engineering, surface properties and structures of engineering materials. He has authored or coauthored more than 300 scientific publications worldwide, including 12 monographs and books, and more than 80 publications in the Philadelphia list. He has won 18 national and international awards and honors, and has served, or is currently serving, as a supervisor or contractor for more than 15 research and didactic projects in Poland and abroad. Prof. Tański is also a reviewer and promoter of numerous scientific papers, including seven doctoral research papers in the field of nanotechnology and materials.

Dr. Wojciech Borek, PhD and MSc in Engineering, is an assistant professor at the Institute of Engineering Materials and Biomaterials at the Silesian University of Technology in Gliwice, Poland, and a member of the association of Polish Cluster of Innovative Forging Technologies "HEFAJSTOS." His scientific interests include materials science, heat treatment, thermomechanical treatment, plastic deformations, and Gleeble simulations; also, he is a specialist in steels, stainless steel, high-manganese austenitic steels, light metal alloys, aluminum, magnesium, and zinc. Dr. Borek is the author and coauthor of approximately 70 scientific publications worldwide, including two book chapters and more than 20 publications in the Philadelphia list. He has won 10 awards and honors, national and international, and has served, or is currently serving, as a contractor of more than six research and didactic projects in Poland and abroad. He is a reviewer of numerous scientific publications and copromoter of two doctoral dissertations.

Dr Mariusz Król, PhD and MSc in Engineering, is an assistant professor at the Institute of Engineering Materials and Biomaterials at the Silesian University of Technology in Gliwice, Poland, and a member of the association of Polish Cluster of Innovative Forging Technologies "HEFAJSTOS". His scientific interests include materials science, additive manufacturing, thermal analysis, i.e. UMSA simulations and dilatometry study, and plastic deformations. Dr Król is also a specialist in light metal alloys. He is head of the Workroom of Metal, Ceramic and Gradient Materials Laser Constitution and the Workroom for Simulation of Metallurgy Processes and Dialometric Tests at the Institute of Engineering Materials and Biomaterials, Silesian University of Technology. He is author and coauthor of more than 80 scientific publications worldwide, including three chapters in books and more than 18 manuscripts in the Philadelphia list. Dr Król has won three national and international awards and honours, and has served, or is currently serving, as a contractor of more than four research and didactic projects in Poland and abroad. He is also a reviewer of numerous scientific publications and copromoter of doctoral dissertations.

Contents

Preface

This book contains seven chapters with original and innovative research studies on issues related to magnesium alloys.

Magnesium, previously used for military purposes, seems to fit perfectly into the requirements of the currently prevailing technology. Low density with appropriate mechanical properties (strength, high operating temperature), good foundry properties (high castability and low shrinkage), vibration damping ability and cost-effectiveness of recycling seem to be an ideal response to market needs. All things considered, magnesium alloys are the perfect material used in various industries, starting from the automotive industry, through sport, electronics up to the space industry and defence.

This book was written by experts in various areas of magnesium science and technology. It gives a general idea of modern advancements in theory and practical purposes of magnesium alloys. The book reports fundamental aspects of corrosion types and details about magnesium alloys designed to work in elevated temperatures and superplastic behaviour. Fundamentals, broad experience, theory as well as complex technological aspects make this work helpful for engineers and scientists from all over the world.

Prof. Tomasz Tański
Head of the Institute
Institute of Engineering Materials and Biomaterials
Silesian University of Technology, Poland

Wojciech Borek
Silesian University of Technology, Poland

Mariusz Król
Silesian University of Technology, Poland

Chapter 1

Introductory Chapter: Magnesium Alloys

Tomasz Tański and Mariusz Król

Additional information is available at the end of the chapter

http://dx.doi.org/10.5772/intechopen.81761

1. Introduction

Today the development of techniques and technologies makes it necessary to look for new design solutions which aim to improve efficiency, product quality and reliability. Before selecting the material, the analysis of requirements is carried out. The demand for using materials with the lowest density and good properties has been increasing for years [1, 2].

Currently trends in the development of techniques tend to reduce the weight of finished products, machines or constructions. This is dictated by ecology and reduced energy consumption. It also results in a growing interest in low-density materials, mainly magnesium and its alloys [1–3].

Magnesium is an element constituting 2.74% of the Earth's crust. It occurs in most countries in the form of various compounds and minerals. The most common ores are dolomites ($MgCO_3 \cdot CaCO_3$) and magnesites ($MgCO_3$). Another mineral used for the magnesium production is carnallite. Magnesium is also included in serpentine. Oceans are a rich source of magnesium, which contain approx. 0.13% of the metal in the salt solution as $MgCl_2$ and $MgSO_4$. In 1808, the electrolysis of molten salts allowed Humphrey Davy, an English chemist, to extract pure active metals for the first time, including magnesium [4–6].

A French chemist Antoine Bussy, on the other hand, developed another method of producing metallic magnesium. He published in Mémoire sur le Radical métallique de la Magnésie (1831) the process of obtaining magnesium which involves heating magnesium chloride ($MgCl_2$) and potassium (K) in the retort [1, 3, 4].

After dissolving potassium chloride (KCl) formed as a result of the reaction, pure magnesium remained in the retort. In the mid-nineteenth century, this method allowed magnesium production in Paris and Manchester which was initially almost entirely devoted to photography. Metallic magnesium was obtained in the form of a wire or powder. In 1852, Robert Bunsen using the Davy's concept constructed a bath for the electrolysis of molten magnesium chloride. Several years later, in 1886, the production of magnesium began in Germany using the modification of the Bunsen's bathtub. This process was jointly undertaken by Chemische Fabrik Griesheim-Elektron and Aluminium and Magnesium Fabrik on an industrial scale in 1896. What is more, between 1914 and 1915, it was the only major magnesium producer in the world [2].

Due to low mechanical properties, the element itself is not used as a construction material. Magnesium alloys, however, are completely different. They present much better mechanical and strength properties taking special care to maintaining their most important low-density advantage through various modifying treatments. The interest in these rates dates back to the beginning of the twentieth century. Classical magnesium alloys are a combination of aluminium, magnesium, manganese and zinc. Magnesium combined with lithium forms ultralight alloys that have many uses. Since it is a reasonable material, it offers great possibilities and is constantly tested at various angles of applications and properties [2–4].

A few decades ago, magnesium alloys were used mainly for military purposes. Today due to the development in plastic deformation, heat treatment and manufacturing (e.g. cooling by high cooling rate, thixocasting), they are used in other fields, such as the automotive or aerospace industries.

In addition to low density, they exhibit many other benefits, such as good corrosion and castability, high dimensional stability and resistance to shocks and impacts, low shrinkage, the ability to be applied to 300°C elements as well as the ease of welding and recycling [4–7].

Magnesium, previously used for military purposes, seems to fit perfectly to the requirements of the currently prevailing technology. Low density, with appropriate mechanical properties (strength, high operating temperature), good foundry properties (high castability and low shrinkage), vibration damping ability, and cost-effectiveness of recycling, seem to be an ideal response to market needs [1–3, 7].

All things considered, magnesium alloys are the perfect material used in various industries starting from the automotive industry, through sport, electronics up to the space industry and defence.

That work was developed by experts in various areas of magnesium science and technology. It gives a general idea of modern advancements in theory and practical purposes of magnesium alloys. The book reports fundamental aspects of corrosion types, details about magnesium alloys designed to work in elevated temperature and superplastic behaviour. Fundamentals, broad experience, theory as well as complex technological aspects make this work helpful for engineers and scientists from all over the world.

Author details

Tomasz Tański* and Mariusz Król

*Address all correspondence to: tomasz.tanski@polsl.pl

Institute of Engineering Materials and Biomaterials at Silesian University of Technology, Poland

References

[1] Friedrich H, Mordike B. Magnesium Technology. Berlin Heidelberg: Springer-Verlag; 2006. DOI: 10.1007/3-540-30812-1

[2] Dobrzanski LA, Tanski T, Malara S, Krol M, Domagala-Dubiel J. Contemporary forming methods of the structure and properties of cast magnesium alloys. In: Czerwinski F, editor. Magnesium Alloys—Design, Processing and Properties. Rijeka, Croatia: Intech; 2011

[3] StJohnl DH, Dahlel AK, Abbott T, Navel MD, Ma Qianl. Solidification of Cast Magnesium Alloys. In: Kaplan HI, editor. Magnesium Technology. The Minerals, Metals & Materials Society; 2003

[4] Mathaudhu SN, Luo AA, Neelameggham NR, Nyberg EA, Sillekens WH. Essential Readings in Magnesium Technology. Hoboken, NJ, USA: John Wiley & Sons, Inc.; 2014. pp. 1-634

[5] Gupta MS, Ling NM. Introduction to Magnesium, Magnesium Alloys, and Magnesium Composites. Wiley; 2011. pp. 1-12

[6] Xianhua C, Yuxiao G, Fusheng P. Research progress in magnesium alloys as functional materials. Rare Metal Materials and Engineering. 2016;**45**(9):2269-2274

[7] Ciach R. Advanced light alloys and composites. Journal of Chemical Information and Modeling. 1989;**53**:160

Chapter 2

A Critical Review of the Application of Electrochemical Techniques for Studying Corrosion of Mg and Mg Alloys: Opportunities and Challenges

Santiago Fajardo, Federico R. García-Galvan, Violeta Barranco, Juan C. Galvan and Sebastian Feliu Batlle

Additional information is available at the end of the chapter

http://dx.doi.org/10.5772/intechopen.79497

Abstract

In order to elucidate the corrosion mechanism of Magnesium (Mg), assess its corrosion rate and evaluate the viability of effective corrosion protection methods, a number of different and complementary techniques are required. Aqueous corrosion is, in nature, an electrochemical process and as such electrochemical methods represent a powerful tool for the study of Mg corrosion. In this chapter the main electrochemical techniques used to study the corrosion of Mg are reviewed along with other simple non-electrochemical methods such as weight loss and hydrogen evolution measurements. The electrochemical techniques covered in this review include conventional DC and AC electrochemical techniques and the latest advances in local electrochemical methods for the evaluation and characterization of Mg corrosion. Each technique presented will be discussed, and its major advantages and drawbacks for the study of Mg corrosion will be commented. Applications range from studies of influence of the impurities in catalytic activity of high purity Mg towards hydrogen evolution, the determination of corrosion rate for Mg and Mg alloys by electrochemical methods and electrochemical study of sol-gel films as pre-treatment for Mg alloys.

Keywords: magnesium alloys, corrosion resistance, hydrogen evolution, anodic dissolution, cathodic reaction, electrochemical impedance spectroscopy (EIS), polarization, local electrochemical impedance spectroscopy (LEIS), sol-gel coatings

1. Introduction

Magnesium (Mg) and Mg alloys are materials of great technological interest. Its low density (approximately 1.7 g cm^{-3}) even lower than that of aluminum, combined with good mechanical properties makes them the lightest structural materials [1]. Mg and its alloys are used for a number of different applications such as biomaterial for biodegradable implants, electrode battery material for primary and secondary Mg batteries and casings for electronic devices, among others. Furthermore, the increasing demand of more environmentally sustainable transports, in particular the lightweighting of vehicles, makes Mg alloys excellent candidates for the automotive sector [1].

Despite the increasing technological possibilities of Mg alloys its use has been limited so far due to its high reactivity in aqueous solutions. Mg exhibits the lowest standard reduction potential among all structural metals (approximately −2.4 V vs. SHE). This results in very high corrosion rates when in contact with water [2].

There are a number of methods to assess the corrosion rate of Mg. As with most metals, the most common are weight loss measurements. This simple and well-established method is based in determining the difference in weight of the specimen under study before and after it is immersed in an aggressive electrolyte for a certain amount of time [3]. However, this method presents some important limitations [4]. The first one is that it only provides the average corrosion rate, not allowing for differences in the oxidation kinetics to be determined. Secondly, it can lead to inaccuracies in corrosion rate determination as a consequence of a poor or an excessive corrosion product removal post immersion.

Another common procedure to assess the corrosion rate of Mg alloys is the hydrogen collection method. In this method the rates associated with the HER in a quiescent electrolyte are determined [5]. The primary cathodic reaction in Mg dissolution in the absence of an external polarization is the hydrogen evolution reaction (HER). This is due to the very low corrosion potential (E_{corr}) exhibited by Mg and its alloys in solution (between −1.3 and − 1.8 V vs. SCE), well below the reversible potential for the HER ($E_{rev,H}$) that creates a large overpotential for the HER [1]. Under these circumstances the oxygen reduction reaction (normally diffusion limited) does not control the cathodic kinetics and the HER dominates the rates associated with the overall reduction process. Given that at the E_{corr} the rate of oxidation is equal to the rate of reduction, the corrosion rate of Mg can be easily assessed by recording the evolution of H_2 volume with the time of immersion. Please note that in this case, in contrast with the weight loss measurements, the Mg dissolution kinetics is determined indirectly by assessing the rates associated with the main cathodic reaction.

There are two main experimental arrangements for measuring the evolution of H_2 from a dissolving Mg specimen, as shown in **Figure 1**. In the traditional setup (see **Figure 1a**) the H_2 gas volume is collected by a burette attached to a funnel that are placed upside down over the sample and filled with electrolyte. As Mg dissolves the evolved H_2 travels upwards to the top of the burette, displacing the solution so the volume of gas can be easily monitored over time [6]. Even though this setup has been common for a long time it presents important limitations [4]. Recently, Fajardo and Frankel developed a more sensitive setup originally proposed by Curioni [7] in which the evolution of hydrogen is monitored by attaching the specimen to a fully

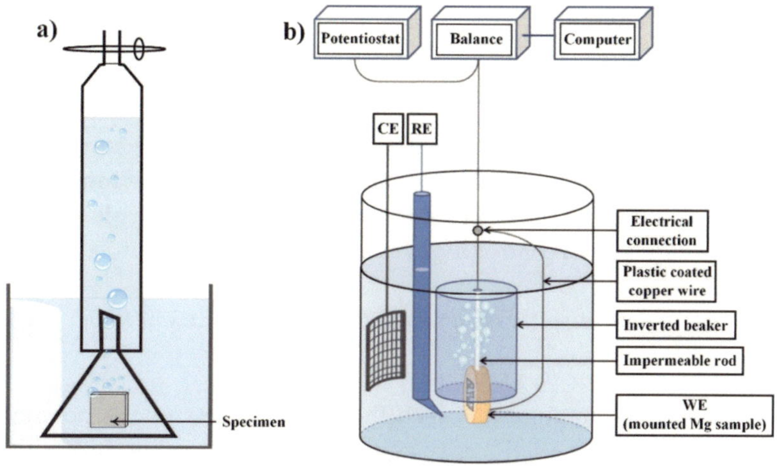

Figure 1. Schematic of the experimental setups for hydrogen gas collection [1]: (a) volumetric method [6], and (b) gravimetric method [5].

immersed container coupled with a high precision balance [5] (see **Figure 1b**). The gas that accumulates at the top of the container changes the apparent weight measured by the balance due to a variation in the buoyant force produced by the volume of solution displaced by the evolved H_2.

Besides weight loss and H_2 volume collection measurements, the electrochemical methods continue to be the most popular in corrosion science and Mg is no exception. This is essentially due to the electrochemical nature of aqueous corrosion. They represent a reliable, rapid and often simple way of assessing many aspects of the corrosion of metals such as kinetics, passivity, etc. Furthermore, the application of electrochemical techniques for the study of Mg corrosion can be used independently or simultaneously with the non-electrochemical methods previously described (among others). In this chapter the main electrochemical techniques that are commonly used in the study of Mg corrosion will be presented. Global electrochemical methods such as potentiodynamic/potentiostatic and galvanodynamic/galvanostatic polarization measurements will be described and critically discussed. Conventional electrochemical impedance spectroscopy (EIS) will be also covered. Furthermore, localized electrochemical methods have shown to be highly useful for the study of many aspects of Mg corrosion and corrosion protection. For this reason, localized electrochemical techniques such as the scanning electrochemical microscopy (SECM), the scanning vibrating electrode technique (SVET), localized electrochemical impedance Spectroscopy (LEIS) and the scanning Kelvin probe (SKP) will be reviewed, among others.

2. Global electrochemical techniques

Mg corrosion in aqueous solutions is in nature an electrochemical process and as such, electrochemical techniques are a powerful tool to assess the corrosion rate of Mg and Mg alloys. In this section, the main global electrochemical techniques used in the study of Mg corrosion, either by the application of a direct current (DC) or an alternating current (AC), will be discussed.

2.1. Direct current (DC) methods

In these electrochemical methods a unidirectional flow of charge is passed through the electrochemical cell either by controlling the voltage and monitoring the current, or vice versa. A three electrode configuration is commonly used. The specimen under study acts as the working electrode, a reference electrode is used to measure the potential of the working electrode, and a counter or auxiliary electrode allows to close the electrical circuit. Finally, a potentiostat/galvanostat is employed to control the electrochemical cell.

2.1.1. Potentiodynamic polarization (PDP) and potentiostatic polarization (PSP) measurements

This is a voltage control technique in which the working electrode is polarized at a fixed rate over a range of potentials (PDP) or remains at a constant value during the entire time of experimentation (PSP). The current flowing through the cell in response to the generated electric field is then recorded. PDP allows to determine the kinetics of the total anodic and cathodic processes when the electrode potential is scanned at potentials above and below the E_{corr}, respectively. This is clearly shown in **Figure 2** where the PDP plots of a Mg specimen with 2 different purities are presented [2]. Furthermore, PDP measurements allow for any passive behavior to be detected. This is defined by the potential region in a V vs. log i plot (V being potential and I being current density) where the current remains constant with the increase in the applied anodic potential.

If the electrochemical system is controlled by activation (i.e. the rate limiting step is the charge transfer reaction), it is possible to determine the corrosion current density (analog to the corrosion rate) by extrapolating the straight lines shown in the cathodic and anodic branches of a PDP plot to the corrosion potential. These lines are called Tafel lines and normally extend through a region in a range of about ±50 to ±250 mV from the E_{corr} [8]. An example of this protocol is shown in **Figure 3**. In this case the Tafel extrapolation was carried out on pure Mg in phosphate-buffered saline (PBS) using a rotating disc electrode (RDE) [9].

Instantaneous corrosion rate can also be determined by the linear polarization resistance (LPR) method [10]. In this case, the electrode potential is scanned over a much smaller range (normally ±10 mV vs. E_{corr}) where a linear-like behavior is expected under activation controlled kinetics [8]. **Figure 4** shows the linear polarization resistance plots of a Mg-Y-RE Mg alloy where the grain size varied in a range from 70 to 0.7 μm [11]. The linear regions expected close to the E_{corr} are clearly shown. The slope of the linear region is defined as the polarization resistance (R_p) and is inversely proportional to the corrosion rate, which can be calculated then using the Stern-Geary equation (see Section 3.2) [8]. The major advantage of the LPR method is that it can be considered as non-destructive given the small polarization applied to the electrode in comparison to an extended PDP measurement where the Tafel regions want to be determined.

However, Tafel-extrapolation and the LPR method are not trivial for Mg due to the increasing ohmic potential drop resulting from the continuously increasing flow of current and resistance associated with the HER (occurring both under anodic and cathodic polarization) [1]. At this point is necessary to comment that Mg and its alloys contradict standard electrochemical kinetics, showing increasing rates of HE with increasing anodic polarization [2].

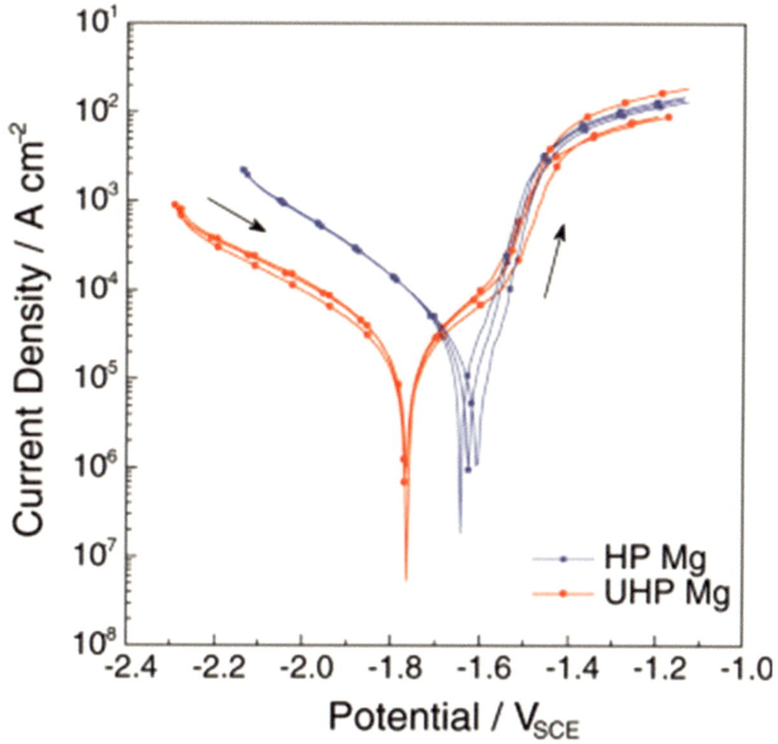

Figure 2. PDP curves of high purity and ultra-high purity (HP and UHP, respectively) Mg in NaCl solution [2].

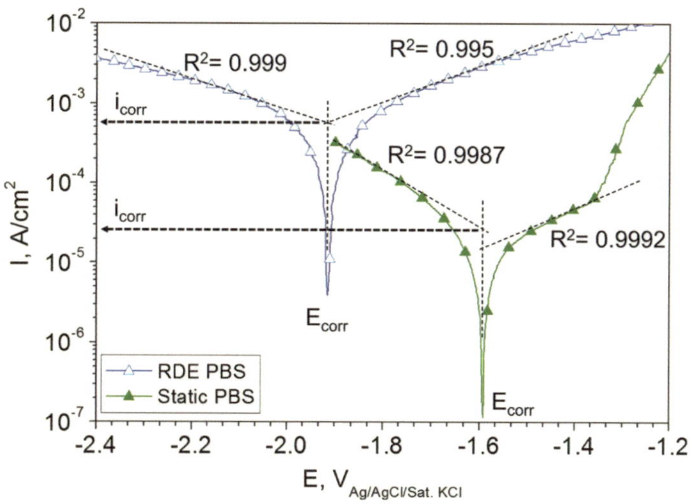

Figure 3. Tafel extrapolation on PDP curves of pure mg under static and rotating conditions in PBS [9].

This phenomenon, historically referred to as negative difference effect (NDE), opposes the expected behavior for an electrochemical system. According to standard electrochemical kinetics (as exemplified by the Butler-Volmer equation) the rate associated with an electrochemical

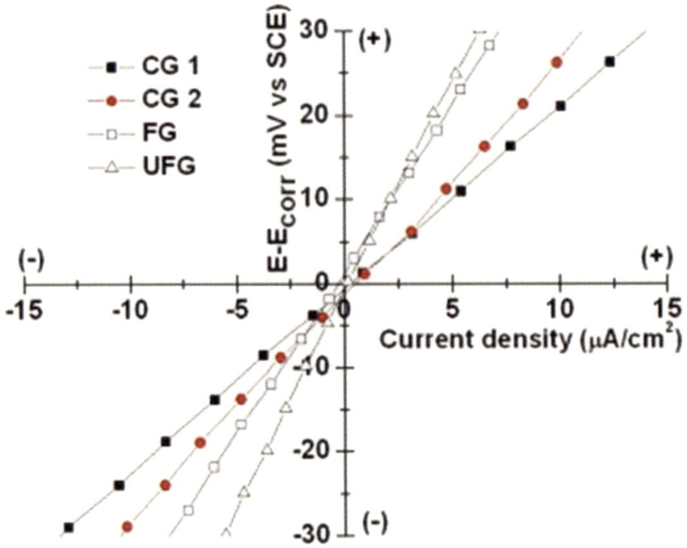

Figure 4. Linear polarization resistance plots of different conditions of Mg–Y–RE alloy in 3.5 wt.% NaCl [11].

half reaction should decrease exponentially with the increase in polarization of the opposite polarity [12]. As a consequence of this anomalous HE, ohmic potential drop effects dominate the shape of the PDP plot, difficulting the identification of the Tafel regions to carry out the extrapolation or distorting the linear-like behavior expected for a LPR measurement [3].

In PSP measurements a constant potential is applied to the working electrode. This method is normally less common than PDP. However, it is useful in some cases like, for example, for the study of the cathodic activation of Mg under anodic polarization [13]. **Figure 5** shows the evolution with time of the measured current density for high purity Mg under PSP at an anodic potential (i.e. more positive than its E_{corr}). As observed, the electrode undergoes a polarity reversal, exhibiting a net cathodic behavior after about 20 min of polarization even if the applied potential is anodic.

2.1.2. Galvanodynamic polarization (GDP) and galvanostatic polarization (GSP) measurements

As opposed to the potential-controlled methods, GDP and GSP measurements are current control techniques. This means that current is either scanned over a certain range or applied at a constant value (i.e. GDP and GSP, respectively) and the potential is measured [8]. Please note that according to standard electrochemical kinetics potential and current are mutually dependent.

Even though galvanodynamic polarization is not a suitable method when the passive behavior of a metallic surface wants to be assessed, GDP and GSP measurements are particularly useful for the study of certain aspects of Mg corrosion. One example is the investigation of the anomalous HE exhibited by anodically polarized Mg. Current control allows for an easy comparison between the current associated with the evolution of H_2 and the applied current density [2, 14, 15]. This has been traditionally evaluated in Mg and Mg alloys using

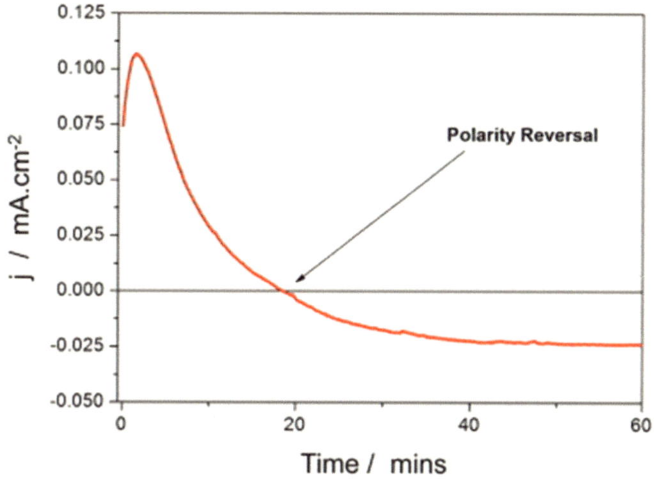

Figure 5. Net current density measured during PSP of high purity Mg at −1.6 V_{SCE} in NaCl solution [13].

GSP, where a constant anodic current density flows through the electrochemical cell and the hydrogen gas volume evolving from the Mg sample is collected. **Figure 6a** shows the volume of H_2 measurements as a function of time determined from gravimetric measurements on ultra-high purity Mg in NaCl solution using GSP under the application of different anodic current densities [14]. From the steady-state rates during GSP the HE current density values can be calculated (see **Figure 6**). The reason for the preferential use of GSP instead of GDP is that it has not been until recently that a hydrogen collection method with a suitable temporal resolution has been available (see Section 1) [5]. Since the development of the gravimetric method for HE, GDP measurements have shown to reliably monitor the real-time evolution of H_2 from a Mg surface. Using GDP instead of GSP allows for an easy determination of the HE current density in a single GDP measurement by interpolation in a HE current density vs. applied current density plot [16] (see **Figure 7**). This has shown to be particularly useful when short timescales are needed or when limited amount of material is available and a large range current densities need to be applied.

Another advantage of using current control methods such as GDP and GSP instead of potential control techniques is that the sample under study can be more reliably polarized before ohmic effects dominate the electrochemical response of the system. Under potential control the large ohmic potential drop associated with the intense HE exhibited both during cathodic and anodic polarization difficult an accurate polarization to a particular potential value. Furthermore, the high reactivity of Mg in aqueous solutions makes it behave like an ideally non-polarizable electrode (i.e. an increase of several orders of magnitude is shown within a small range of applied potentials above the E_{corr}), limiting the region over which the electrode can be effectively polarized. Consequently, if current is controlled instead of potential it is possible to accurately expand the study of the electrochemical behavior of Mg and Mg alloys over a wider range of polarization.

It is also interesting to mention that, due to the anomalous HE exhibited by Mg during anodic polarization, the current densities registered by the potentiostat do not resemble the true dissolution kinetics of Mg. This evolution of hydrogen is a cathodic process and as such, consumes

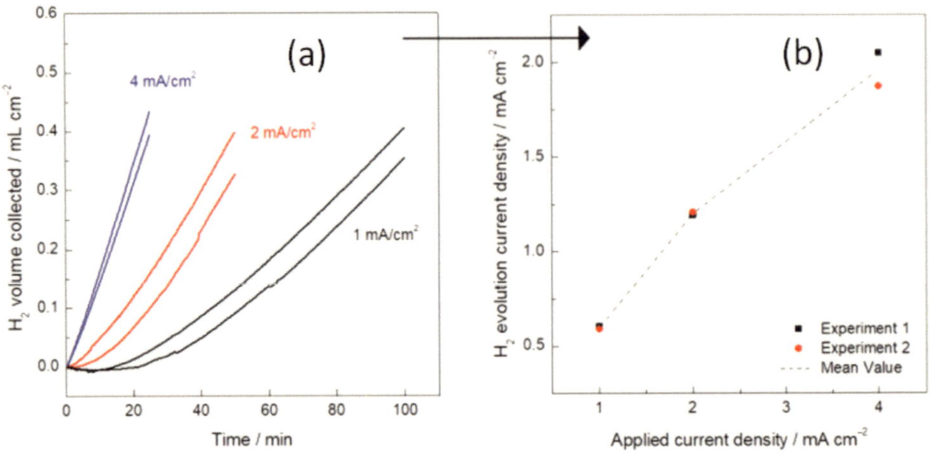

Figure 6. (a) Volume of H_2 determined from gravimetric measurements as a function of time for an ultra-high purity Mg electrode in NaCl solution during a GSP measurement and (b) their corresponding HE current density values (right) [14].

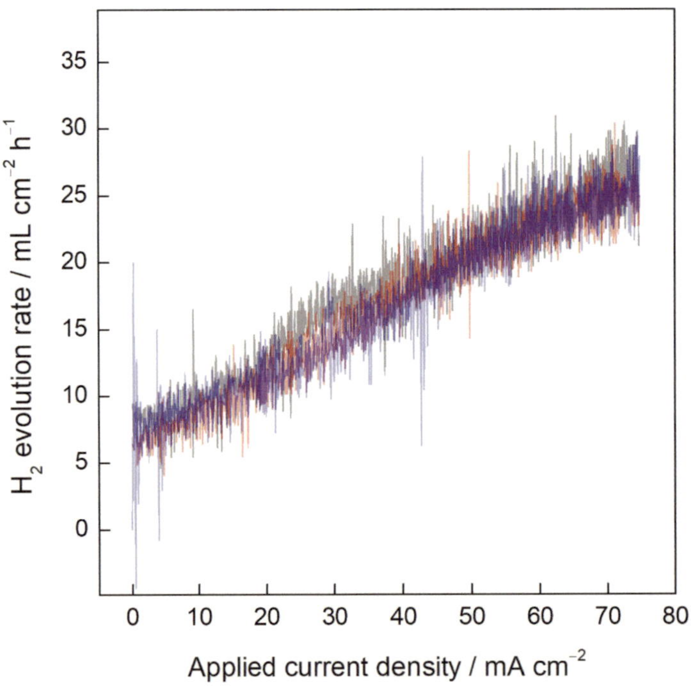

Figure 7. H_2 evolution rates for a high purity Mg electrode in 0.1 M citric acid buffer (pH 3) solution during anodic GDP as a function of the applied current density [16].

electrons. The source of electrons on the anodically polarized surface is the Mg specimen. Please note that these electrons are consumed at the electrode surface and do not flow through the potentiostat. Consequently, the Mg extra dissolution associated with the anomalous HE is not accounted in the current density measured in a simple polarization experiment. To accurately assess the true Mg dissolution kinetics, it is necessary to simultaneously account

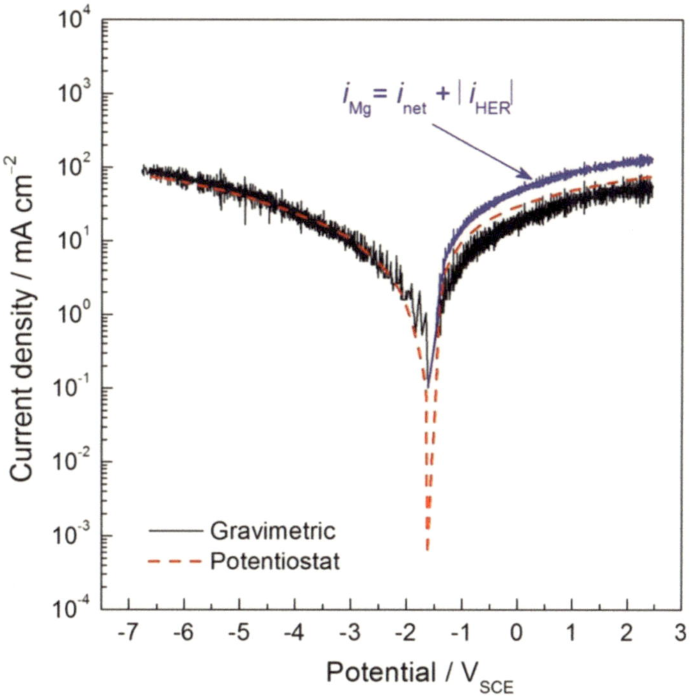

Figure 8. GDP curve of a high purity Mg electrode in NaCl solution as a function of the measured potential [5].

for the real-time HE occurring during polarization. This way, the true current density associated with Mg oxidation will be given by the sum of the net current density measured by the potentiostat (electrons that flow through the electrochemical cell) and the HE current density (consequence of the anomalous HE and measured using a suitable hydrogen collection method) [5] (see **Figure 8**).

In summary, direct current electrochemical methods are a powerful technique to evaluate the electrochemical behavior of Mg and its alloys. They allow for the reaction kinetics and corrosion rate to be determined as well as to assess passivity. Furthermore, they are useful when the evolution of hydrogen is investigated. However, it is necessary to be particularly careful with the ohmic potential drop that is difficult to compensate since they may distort the shape of a polarization curve leading to inaccurate results.

2.2. Electrochemical impedance spectroscopy (EIS) technique

EIS is one of the most widely used technique for investigating in situ the corrosion mechanisms and surface films developed on Mg and alloys specimens in corrosive environments [1, 3, 4, 17–20]. EIS studies are also useful to evaluate the corrosion rate of Mg [1, 3, 17] and to rank the corrosion protective ability of Mg alloys [20].

2.2.1. Advantages

Its most outstanding and well-known advantages are its relative ease of use, the relatively short time taken for measurements, the use of relatively cheap and simple equipment, quantitative nature, high accuracy and reproducibility of the results [18]. Apart from the quantitative information provided by this technique, the occurrence of corrosion in Mg alloys can also be indicated by the emerging of EIS inductive characteristics long before visual changes can be observed using traditional exposure tests [21].

While polarization curve technique is destructive to the specimen and cannot serve for prediction of the long-term corrosion rates of the material [4], the surface properties of the metal or metal alloy remain similar after each EIS measurement since signals applied have low amplitude (the deviation from corrosion equilibrium was ±5 to ±10 mV) and do not alter corrosion potential [18, 21]. The non-destructive character of the impedance technique [19] allows continuous monitoring of the progress of corrosion process in situ, with instantaneous measurements of the corrosion rate and can provide information on the changes of the mechanisms of degradation for the Mg alloys during the immersion period [17–19]. EIS has been considered as an ideal method to evaluate the durability of these alloys [20].

2.2.2. *Limitations*

A major shortcoming with conventional EIS measurements is associated with the lack of spatial resolution, as the measured impedance results are attributed to the electrochemical response of the global properties of the corroding system, reflecting an averaged electrochemical behavior of the whole electrode [22]. This is a major limitation in the investigation of localized corrosion processes such as passivity breakdown, pitting corrosion and the breakdown of coated systems [22].

Equivalent circuit models using passive electrical engineering and physics circuit elements have commonly been used in order to convert frequency response data to corrosion properties (e.g. resistance and impedance) [4]. However, one of the major weaknesses of this analytical approach arises in how to select the equivalent circuits that not only fit the values, as often multiple equivalent circuits may fit the same data [4], but also provide a meaningful interpretation of the studied interface with quantitative parameters [4].

2.2.3. *Typical spectra and equivalent circuits*

In our EIS studies about the corrosion of Mg alloys, we reported Nyquist plots characterized by two loops (**Figure 9a**), a capacitive loop at high frequencies (HFs) and an inductive loop at low frequencies (LFs) [18] which can also be observed in corresponding Bode plots (**Figure 9b** and **c**). It is generally agreed that the capacitive loop at high frequencies is always related to the charge transfer resistance as well as the effect of ionic double layer capacitance of the electrode. The equivalent circuit shown in **Figure 10a** was used to simulate the two-time constant impedance spectra in **Figure 9a**. In this circuit, Rs represents the electrolyte resistance, R1 is the charge transfer resistance, and CPE1 represents the non-ideal capacitive behavior related to the electrical double layer [18]. An inductor (L) and a resistance (RL) have been included to represent the inductive response appearing at low frequency [17–20].

Most of EIS studies about the corrosion of Mg alloys reported impedance diagrams rather complex, exhibiting three loops (**Figure 9d**), a capacitive arc in the high frequency (HFs) region, a capacitive arc in the middle frequency (MFs) region, and one inductive loop in the low frequency (LFs) region [23–26]. This behavior can be confirmed using Bode plots

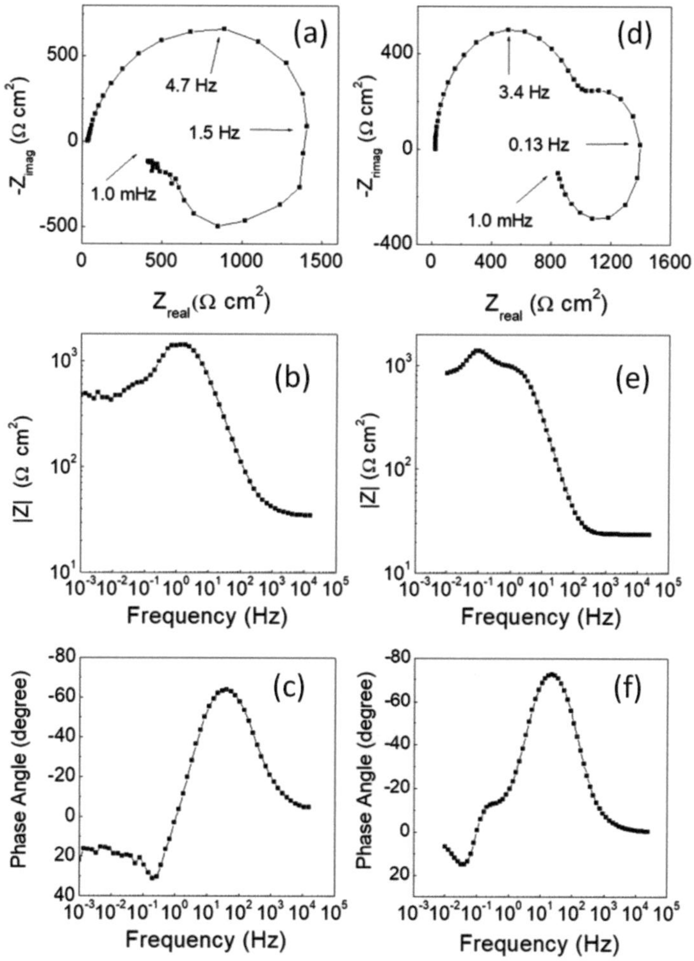

Figure 9. (a, d) Typical Nyquist and (b, c, e, f) bode plots for Mg alloys.

(**Figure 9e** and **f**). In this case, an additional R2/CPE2 model element was introduced in the equivalent circuit (**Figure 10b**). Some studies suggested that the capacitive loop in the HFs may be ascribed to the charge transfer process at the double layer formed at the surface film while the capacitive loop in the MFs accounts for mass transport processes [21, 24, 25]. However, the interpretation of the two characteristic capacitive loops in the high and medium frequency ranges is controversial and unclear [23, 26].

2.2.4. Challenges

Stern and Geary [10] shown that the that the polarization resistance (R_p) is inversely proportional to the corrosion current density in cases when metals behave as electrochemically active

$$i_{corr} = B/R_p \tag{1}$$

where B is the Stern Geary coefficient that is a function of the cathodic and anodic Tafel slopes (β_c and β_a) [17, 19].

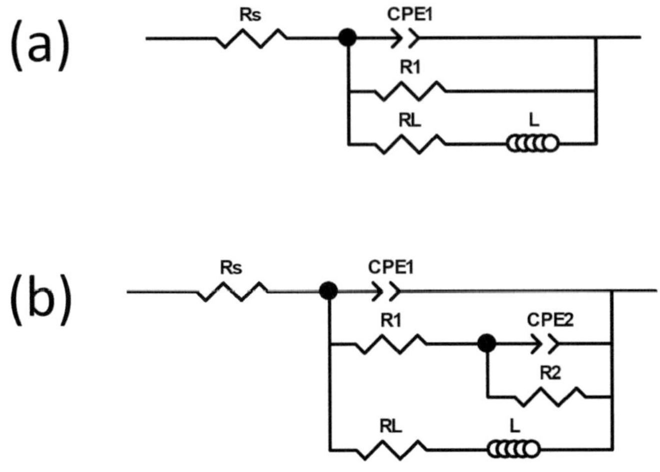

Figure 10. (a) Equivalent circuit used for fitting two-time and (b) three-time constants impedance spectra.

The use of the Stern-Geary equation (1) for the quantitative determination of the corrosion rate for Mg and its alloys presents two main challenges: the precise knowledge of the B constant and the value of resistance obtained from fitting of the EIS spectra selected for the estimation of i_{corr}.

As already commented (**Figure 9a** and **d**), the impedance diagrams for Mg usually are characterized by a well-marked inductive loop at LF [17]. The inductive behavior is often associated with the occurrence of pitting corrosion caused by anodic dissolution of the Mg and linked to the relaxation of adsorbed intermediates [27]. The disappearance of the inductive loop is caused by the formation of protective corrosion layers on Mg surface which inhibit the pitting corrosion [28]. However, the precise chemical species responsible for the characteristic inductive loop are still unclear [17] and its interpretation remains controversial [19].

2.2.5. Using EIS to monitor corrosion rate

The Nyquist diagrams of AZ31 specimens in NaCl 0.6 M solution display one semi-circle followed by an inductive loop at lower frequency region (**Figure 11a**). After the first hour of immersion in 0.6 M NaCl solution, the charge transfer resistance values, R_t, increased markedly, as shown in **Figure 11b** related to the formation of a corrosion layer with major protective ability. Finally, **Figure 11c** compares the variations of EIS-estimated corrosion rate with immersion time obtained by integration of the current density (i_{corr}) vs. time data determined by the evolution of reciprocal R_t over time and the B′ values calculated via correlation with gravimetric measurements (**Table 1**). It should be noted that corrosion rates determined from parameter R_t quantitatively agree with the corrosion rate calculated by using independent hydrogen evolution measurement over the whole duration of the measurement (**Figure 11c**).

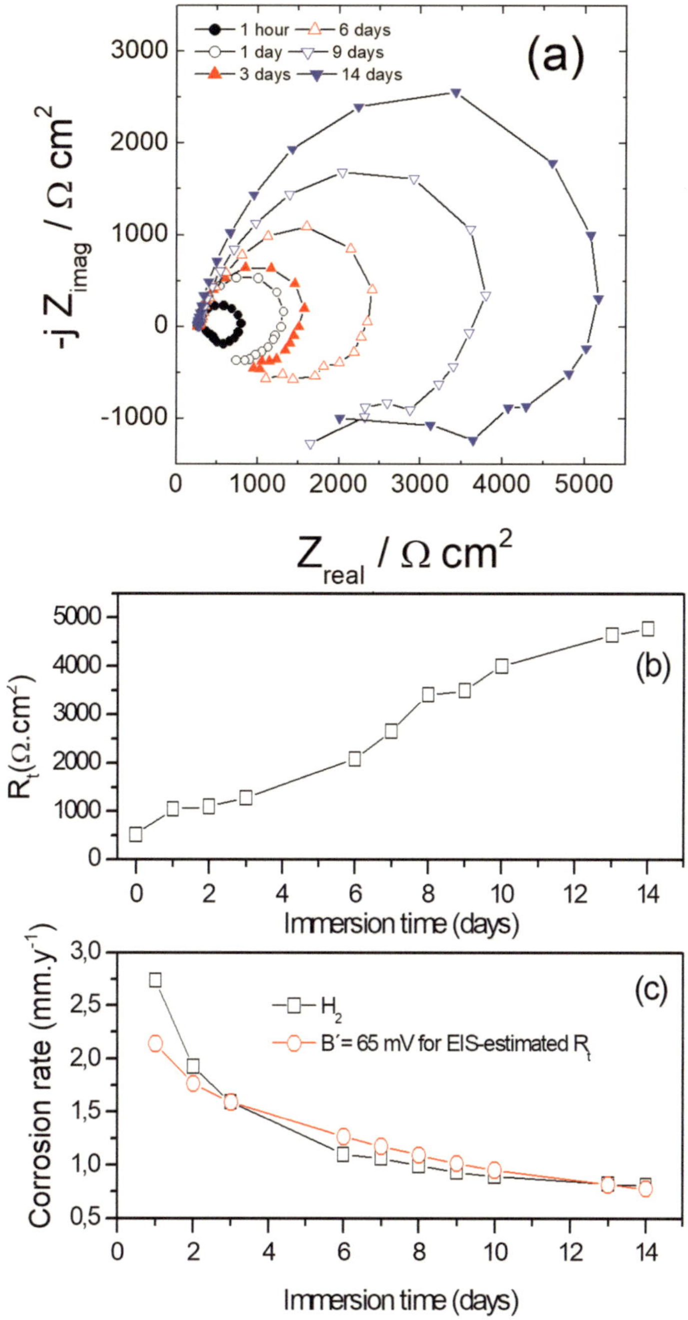

Figure 11. (a) Variation in Nyquist plots for AZ31 sample with immersion time in 0.6 M NaCl solution; (b) evolution of the resistance (R_t) value with immersion time; and (c) variation in corrosion rate (mm.year^{-1}) as a function of immersion time obtained from hydrogen evolution measurements and EIS during 14 days of immersion in 0.6 M NaCl.

Time (days)	$\int(1/R_t)\,dt$	P_w (mm.y^{-1})	B' (mV)
14	7.35.10^{-3}	0.78	65

Table 1. 'Apparent' Stern-Geary coefficient calculated from R_t resistance performed in 0.6 M NaCl solutions.

3. Localized electrochemical techniques for studying Mg and Mg alloys

Corroding metallic surfaces present heterogeneous surfaces and that influence their electrochemical activity, including highly reactive sites and/or passive regions. This is the case of Mg and Mg alloys. So, in order to be able to study the different processes that take place on an Mg corroding surface at the micrometer scale (i.e. corrosion initiation, corrosion propagation, corrosion inhibition by smart coatings, etc.) the use of localized electrochemical techniques can provide crucial information not available using solely global electrochemical methods.

Mg, as previously mentioned, is a very reactive metal that readily reacts with water, producing large amounts of H_2 gas. This H_2 evolution leads to a change in the natural convection of the electrolyte and thus in the electrochemistry of the system under study. The exposition time to the aggressive medium is another important factor, since the increase in pH of the solutions, due to the corrosion processes, causes the precipitation of $Mg(OH)_2$, leading to a further change in the electrochemical activity of the material. In this section the advantages and challenges faced by localized electrochemical techniques to characterize Mg corrosion, will be highlighted, as this remains a complex task.

At present, two different methodologies are applied for local electrochemical investigation:

a. Scanning techniques, where the whole metal surface is either immersed in aqueous electrolyte [Scanning Electrochemical Microscopy (SECM), Scanning Vibrating Electrode Technique (SVET), Localized Electrochemical Impedance Spectroscopy (LEIS)] or exposed at high humid atmosphere [Scanning Kelvin Probe (SKP)].

b. Methodologies where only small areas are exposed to the electrolyte using mini or micro capillary electrochemical cells [Scanning Droplet Cell (SDC), Flowing electrolyte type SDC (FT-SDC), Mini Cell System (MCS)].

The scanning techniques can perform a relatively rapid mapping of the studied areas but, generally, highly diluted electrolytes with low conductivity are required in order to achieve good local resolution. Furthermore, the scanned area cannot be locally polarized. The methods of the second group, on the other hand, have the main advantage that concentrate electrolytes can be used and the current resolution is improved down to the pA to fA range, allowing the study of local dissolution processes in the nm-range. Moreover, any DC or AC electrochemical technique can be used to study the surface with localized resolution. The main disadvantage of these methods is, in general, the slow mapping process. Furthermore, the surface conductivity of the studied metallic material plays an important role in the local resolution achieved. Let's continue firstly, showing the applications of these capillary electrochemical techniques and secondly those of the scanning techniques.

3.1. Microcapillary electrochemical techniques

These techniques basically consist in the miniaturization of a three electrode electrochemical cell. The SDC the electrochemical cell is formed by a micro reference electrode, micro counter electrode and a microcapillary filled with the electrolyte solution that will be in contact with a microarea of the working electrode, which is the metallic surface [29]. In certain designs, which are the case of the FT-SDC, the electrolyte solution can also flow through the microcapillary, allowing fresh solution to be brought to the substrate surface. In new configurations of the scanning droplet microscope (FT-SDCM) [30] the gas bubbles or reaction products formed on the surface of the working electrode can be removed using a constant electrolyte flow, which is extremely useful for the study of Mg and its alloys. This represents an improvement with respect to previous experimental setups, because the cell can operate with stagnant or flowing electrolyte. In the study of the heterogeneous microstructured AZ91 alloy, potentiodynamic polarization scans were recorded using microcapillaries of 5–10 μm diameter [31]. The scans recorded at lowest current were associated with the β phase, meaning that this phase was the most corrosion resistant. These results highlight the suitability of the SDC technique for monitoring the electrochemical activity of microstructural feature avoiding the interference from the bulk of the alloy. For the study of the new smart coatings developed for Mg corrosion protection, it is necessary increase the capillary diameter to reduce the big resistances (MΩ) provided by the coatings when measured in areas of 5–10 μm diameter. The Mini Cell System (MCS) with a geometry pencil like incorporates a three electrode cell with a capillary diameter of 700 μm which allows separate the kinetics of damaged and undamaged areas of a metal/coating system. The MCS has been proved to be extremely useful to characterize new electrochemically active sol-gel thin films applied onto metallic substrates [32, 33].

3.2. Scanning electrochemical techniques

The Scanning Kelvin Probe (SKP) allows the study of localized corrosion processes underneath insulating coatings [34] relevant for the corrosion processes at the polymer-coated metal interface, like swelling and ion incorporation into the coating [35]; the initiation of localized corrosion in defects; and corrosion propagation, caused by anodic or cathodic reactions occurring on the underlying interface. The SKP measures the work-function of a sample using the vibrating condenser method. The major advantage of the SKP in comparison to conventional electrochemical devices is the fact that the SKP measures electrode potentials without touching the surface under investigation across a dielectric medium of high resistance. The measurements are performed in humid atmosphere but in absence of aqueous electrolyte.

The Scanning Electrochemical Microscopy (SECM) uses an ultra-microelectrode (UME) immersed in a liquid electrolyte that bathes the entire metal surface. The UME scans the substrate surface at a fixed height detecting the electrochemical activity on the electrode surface as a consequence of the electrochemical response that the probe experiences when in close proximity with the substrate surface measuring the diffusion-limited current of a redox-active couple (e.g., Fe^{2+}/Fe^{3+}). With a spatial resolution in the micrometer scale or below allows the discrimination between the activities of different electroactive species by polarizing the UME at a potential of interest [36]. The SECM has shown to be a powerful technique in the study of Mg alloys in simulated biological fluids [37] and the localized corrosion of Mg alloys in NaCl solutions [38, 39] and has been also used to study the phenomenon of anomalous HE [40].

Another non-destructive technique is the Scanning Vibrating Electrode Technique (SVET) that measures the electric field generated by a sample immersed in a solution using a vibrating probe with a fixed amplitude, frequency and height. The scanning vibrating probe detects the potential gradients ohmically produced by local currents originated from actively corroding surfaces immersed in an electrolyte. Scanning is done with a 3D micro-stepping motorized micro-manipulator [41]. The probe vibration is controlled by a piezoceramic displacement device allowing vibration amplitudes from 1 to 30 μm (perpendicular to the sample surface). Calibration then enables quantification of localized cathodic and anodic activity [41, 42].

Scanning Reference Electrode Techniques have increasing their use over the last 30 years in studying the phenomena associated to the localized metallic corrosion. The principles of SVET measurement have been extensively addressed in pioneer works written by Isaacs [43–45]. A basic equation used by Bastos et al. [41, 40], for calculating the current density in solution, i, and describing the operating principle of SVET is

$$i = \kappa E = -\kappa\, \Delta V/\Delta r \quad (2)$$

which is a form of Ohm's law, where κ is the solution conductivity, E is the electric field in solution and V is the potential difference between two points separated by the distance r in the direction of current flow.

Several examples of the use of SVET in studies of Mg and Mg alloys can be found in renowned works of Williams and McMurray [46]. They have used SVET for studying, for instance, the inhibition of localized corrosion occurring on unpolarized Mg samples immersed in uninhibited aqueous sodium chloride electrolyte [42]. These authors also have used SVET for studying the behavior of a range of potential anionic inhibitors, selected on the basis of their ability to form insoluble precipitates with aqueous Mg^{2+} ions [46]. Their results using SVET for investigating the source of hydrogen evolution from high purity Mg anodically polarized are also significant [14, 16, 47].

Other successful examples of the use of SVET in corrosion studies of Mg alloys are found in several works of Montemor and Ferreira. These authors studied the relation with corrosion

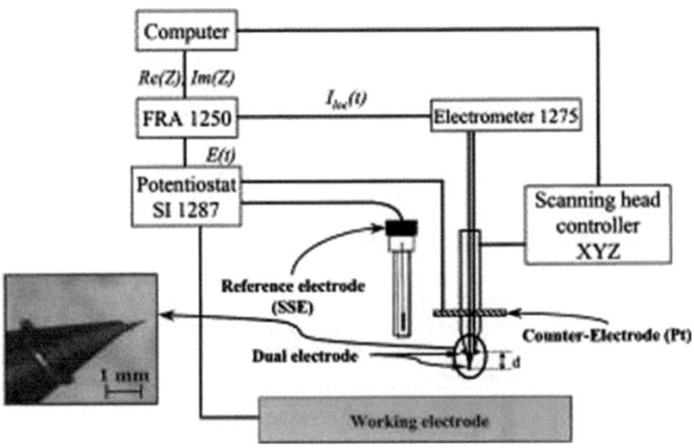

Figure 12. Schematic representation of the LEIS apparatus [56].

protection of pre-treatments consisting in the application of cerium and lanthanum nitrate solutions on AZ31 Mg alloy substrates [48]. Very important contributions of these authors have been their studies with SVET of modified bis-[triethoxysilylpropyl] tetrasulfide silane films on AZ31 alloy and, inhibitor-doped sol-gel coatings (8-Hydroxyquinoline) also on Mg alloy AZ31 substrates [49, 50].

Finally, it is important not to overlook in this section the contributions of the Bierwagen group on the application of SVET to studies of cathodic corrosion protection performance of Mg-rich coatings on aluminum substrates. In such studies coating characteristics were assessed not just using SVET but also multiple electrochemical techniques as SECM, open circuit potential (OCP) measurement, potentiodynamic polarization measurement (PDP) and EIS [51, 52].

Isaacs and Kendig were pioneers in the development of the local electrochemical impedance spectroscopy (LEIS) [53, 54]. This technique is closely related with SVET [55]. Typically, the LEIS method used a five-electrode configuration (**Figure 12**) [56]; a typical three-electrode arrangement (working electrode, counter electrode, and reference electrode) was used to control the dc potential and excite the interface potentiostatically with an ac signal, while two microreference electrodes were used to detect the local potential gradient in solution above the sample surface [57]. The principles of the technique are similar to those used in the conventional bulk EIS; a small sinusoidal voltage perturbation is applied to a working electrode sample and the resulting current is measured to allow the calculation of the impedance. The applied voltage ($\Delta V_{applied}$) is the potential difference between the working electrode and the reference electrode. The local impedance Z_{local} is calculated by the relationship:

$$Z_{local} = \Delta V_{applied}/i_{local} \quad (3)$$

The local AC current density (i_{local}) is calculated using the Ohm's law:

$$i_{local} = (\Delta V_{local}/d)\kappa \quad (4)$$

where ΔV_{local} is the ac potential difference measured between the two probes positioned on and in a conical plastic holder which are separated a distance d, and κ is the conductivity of the electrolyte [56].

Two measurements modes are possible:

- Full local impedance spectrum at one single location (LEIS)
- Area maps of the local impedance of the sample at one frequency (LEIM)

Baril, Galicia and co-workers applied the first measurement mode (LEIS) to study corrosion behavior of pure Mg and AZ91 magnesium Alloy [24, 56, 58]. More recently Barranco, Galvan et al. have applied the second mode (LEIM) to study the corrosion behavior of sol-gel thin films modified with Zr^{4+} and Ce^{3+} ions on AZ9, and doped with organic corrosion inhibitors (benzotriazole and L-cysteine) on AZ31 and AZ61 magnesium alloys [59, 60]. **Figure 13** shows as an example impedance maps (LEIM) recorded around an artificial defect at a fixed frequency (500 Hz) and variable soaking time in 0.006 M NaCl solution for (a) undoped sol-gel film and (b) 0.3% wt.% benzotriazole (BTA) doped sol-gel film deposited on AZ31

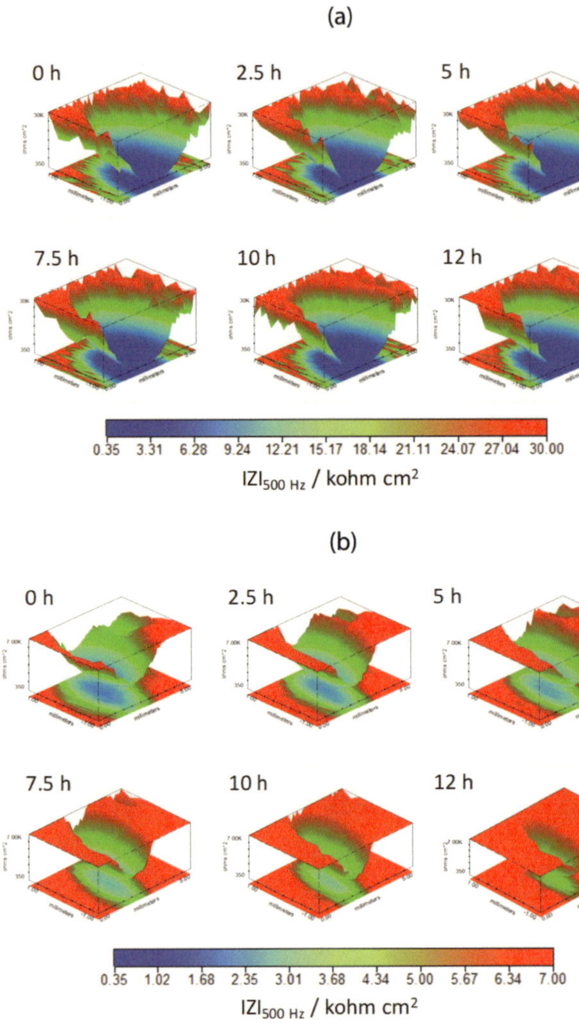

Figure 13. Impedance maps recorded at constant frequency (500 Hz) and variable soaking time in 0.006 M NaCl solution for (a) undoped sol-gel film and (b) 0.3% wt.% BTA doped sol-gel film deposited on AZ31 alloy.

alloy [60]. For the undoped sol-gel films no remarkable differences were observed between the registered impedance maps obtained initially and after longer soak time. In contrast, for BTA doped sol-gel films the LEIS maps showed that around the artificial defect the impedance values increased with soaking time. This effect was ascribed to the gradual release of the BTA from the sol-gel film that caused inhibition of corrosion on the active areas of the magnesium substrate [60].

4. Conclusions

- Global DC methods, comprising potential and current controlled techniques, are versatile tools to assess the corrosion rate of Mg and Mg alloys. They provide relevant information on the reaction kinetics and passive behavior of these materials. Furthermore, DC methods are particularly suitable for the study of the anomalous hydrogen evolution on anodically polarized Mg when coupled with other non-electrochemical methods such as hydrogen volume collection. The main limitation of the global DC techniques for the study of Mg is that, as a consequence of the intense evolution of hydrogen gas (both under anodic and cathodic polarization) large ohmic potential drops normally dominate the electrochemical response not far from the E_{corr}. For this reason, care should be taken when designing, performing and analyzing the results obtained using these methods.

- EIS is a very powerful tool, non-destructive and sensitive electrochemical technique, to study corrosion processes over time, formation of corrosion products and monitor corrosion rates of Mg alloys. These measurements obtained in a very wide range of frequencies allow us to examine the different processes that determine the corrosion behavior of Mg alloys, from the faster processes that are under charge transfer control, to the slower ones, generally of mass transport control, and inductive characteristics due to adsorption species.
- Localized electrochemical techniques, either that based on surface scanning; LEIS, SVET, SKP and SECM or those based on microcapillaries; SDC, FT-SDC, MCS, have proven to be valuable tools in the study of microareas with different electrochemical activity that exist in Mg alloys due to their microstructure.
- Complementary studies based on multiple global and localized electrochemical techniques are the key to face the challenge of controlling the corrosion of reactive Mg and its alloys for their use in technological applications.

Acknowledgements

The authors gratefully acknowledge the financial support provided by the Ministry of Economy and Competitiveness of Spain (MAT2015-65445-C2-1-R) to carry out the present study. Santiago Fajardo expresses his gratitude to the State Research Agency (Ministry of Economy, Industry and Competitiveness of Spain), the Spanish Research Council (CSIC) and the European Regional Development Fund (ERDF) for the financial support under the Project MAT2015-74420-JIN (AEI/FEDER/UE).

Author details

Santiago Fajardo, Federico R. García-Galvan, Violeta Barranco, Juan C. Galvan and Sebastian Feliu Batlle*

*Address all correspondence to: sfeliu@cenim.csic.es

National Centre for Metallurgical Research (CENIM-CSIC), Madrid, Spain

References

[1] Esmaily M, Svensson JE, Fajardo S, Birbilis N, Frankel GS, Virtanen S, Arrabal R, Thomas S, Johansson LG. Fundamentals and advances in magnesium alloy corrosion. Progress in Materials Science. 2017;**89**:92-193. DOI: 10.1016/j.pmatsci.2017.04.011

[2] Fajardo S, Frankel GS. Effect of impurities on the enhanced catalytic activity for hydrogen evolution in high purity magnesium. Electrochimica Acta. 2015;**165**:255-267. DOI: 10.1016/j.electacta.2015.03.021

[3] Bland LG, King AD, Birbilis N, Scully JR. Assessing the corrosion of commercially pure magnesium and commercial AZ31B by electrochemical impedance, mass-loss, hydrogen collection, and inductively coupled plasma optical emission spectrometry solution analysis. Corrosion. 2015;**71**:128-145. DOI: 10.5006/1419

[4] Kirkland NT, Birbilis N, Staiger MP. Assessing the corrosion of biodegradable magnesium implants: A critical review of current methodologies and their limitations. Acta Biomaterialia. 2012;**8**:925-936. DOI: 10.1016/j.actbio.2011.11.014

[5] Fajardo S, Frankel GS. Gravimetric method for hydrogen evolution measurements on dissolving magnesium. Journal of The Electrochemical Society. 2015;**162**:C693-C701. DOI: 10.1149/2.0241514jes

[6] Song G, Atrens A, StJohn D. An hydrogen evolution method for the estimation of the corrosion rate of magnesium alloys. In: Hryn JN, editor. Magnesium Technology. Hoboken, N.J: John Wiley & Sons; 2001. pp. 254-262. DOI: 10.1002/9781118805497.ch44

[7] Curioni M. The behaviour of magnesium during free corrosion and potentiodynamic polarization investigated by real-time hydrogen measurement and optical imaging. Electrochimica Acta. 2014;**120**:284-292. DOI: 10.1016/j.electacta.2013.12.109

[8] Frankel GS. Electrochemical techniques in corrosion: Status, limitations, and needs. Journal of Testing and Evaluation. 2014;**42**:517-540. DOI: 10.1520/JTE20140289

[9] Marco I, Van der Biest O. Polarization measurements from a rotating disc electrode for characterization of magnesium corrosion. Corrosion Science. 2016;**102**:384-393. DOI: 10.1016/j.corsci.2015.10.031

[10] Stern M, Geary AL. Electrochemical polarization: I. A theoretical analysis of the shape of polarization curves. Journal of The Electrochemical Society. 1957;**104**:56-63. DOI: 10.1149/1.2428496

[11] Argade GR, Panigrahi SK, Mishra RS. Effects of grain size on the corrosion resistance of wrought magnesium alloys containing neodymium. Corrosion Science. 2012;**58**:145-151. DOI: 1 0.1016/j.corsci.2012.01.021

[12] Bockris JOM, Reddy AKN. Modern Electrochemistry. New York: Kluwer Academic/ Plenum Publishing; 2000. p. 1432. DOI: 10.1007/978-1-4613-4560-2

[13] Birbilis N, King AD, Thomas S, Frankel GS, Scully JR. Evidence for enhanced catalytic activity of magnesium arising from anodic dissolution. Electrochimica Acta. 2014;**132**:277-283. DOI: 10.1016/j.electacta.2014.03.133

[14] Fajardo S, Glover CF, Williams G, Frankel GS. The source of anodic hydrogen evolution on ultra high purity magnesium. Electrochimica Acta. 2016;**212**:510-521. DOI: 10.1016/j.electacta.2016.07.018

[15] Frankel GS, Fajardo S, Lynch BM. Introductory lecture on corrosion chemistry: A focus on anodic hydrogen evolution on Al and Mg. Faraday Discussions. 2015;**180**:11-33. DOI: 10.1039/C5FD00066A

[16] Fajardo S, Glover CF, Williams G, Frankel GS. The evolution of anodic hydrogen on high purity magnesium in acidic buffer solution. Corrosion. 2017;**73**:482-493. DOI: 10.5006/2247

[17] King AD, Birbilis N, Scully JR. Accurate electrochemical measurement of magnesium corrosion rates; a combined impedance, mass-loss and hydrogen collection study. Electrochimica Acta. 2014;**121**:394-406. DOI: 10.1016/j.electacta.2013.12.124

[18] Feliu S, García-Galvan FR, Llorente I, Diaz L, Simancas J. Influence of hydrogen bubbles adhering to the exposed surface on the corrosion rate of magnesium alloys AZ31 and AZ61 in sodium chloride solution. Materials and Corrosion. 2017;**68**:651-663. DOI: 10.1002/maco.201609233

[19] Curioni M, Scenini F, Monetta T, Bellucci F. Correlation between electrochemical impedance measurements and corrosion rate of magnesium investigated by real-time hydrogen measurement and optical imaging. Electrochimica Acta. 2015;**166**:372-384. DOI: 10.1016/j.electacta.2015.03.050

[20] Jamesh M, Kumar S, Sankara Narayanan TSN. Corrosion behavior of commercially pure Mg and ZM21 Mg alloy in Ringer's solution – Long term evaluation by EIS. Corrosion Science. 2011;**53**:645-654. DOI: 10.1016/j.corsci.2010.10.011

[21] Song G-L, Shi Z. Corrosion mechanism and evaluation of anodized magnesium alloys. Corrosion Science. 2014;**85**:126-140. DOI: 10.1016/j.corsci.2014.04.008

[22] Moreto JA, Marino CEB, Bose Filho WW, Rocha LA, Fernandes JCS. SVET, SKP and EIS study of the corrosion behaviour of high strength Al and Al–Li alloys used in aircraft fabrication. Corrosion Science. 2014;**84**:30-41. DOI: 10.1016/j.corsci.2014.03.001

[23] Pinto R, Ferreira MGS, Carmezim MJ, Montemor MF. The corrosion behaviour of rare-earth containing magnesium alloys in borate buffer solution. Electrochimica Acta. 2011;**56**:1535-1545. DOI: 10.1016/j.electacta.2010.09.081

[24] Galicia G, Pébère N, Tribollet B, Vivier V. Local and global electrochemical impedances applied to the corrosion behaviour of an AZ91 magnesium alloy. Corrosion Science. 2009;**51**:1789-1794. DOI: 10.1016/j.corsci.2009.05.005

[25] Baril G, Pébère N. The corrosion of pure magnesium in aerated and deaerated sodium sulphate solutions. Corrosion Science. 2001;**43**:471-484. DOI: 10.1016/S0010-938X(00)00095-0

[26] Cao F, Shi Z, Hofstetter J, Uggowitzer PJ, Song G, Liu M, Atrens A. Corrosion of ultra-high-purity Mg in 3.5% NaCl solution saturated with $Mg(OH)_2$. Corrosion Science. 2013;**75**:78-99. DOI: 10.1016/j.corsci.2013.05.018

[27] Brett CMA, Dias L, Trindade B, Fischer R, Mies S. Characterisation by EIS of ternary Mg alloys synthesised by mechanical alloying. Electrochimica Acta. 2006;**51**:1752-1760. DOI: 10.1016/j.electacta.2005.02.124

[28] Baril G, Blanc C, Pébère N. AC impedance spectroscopy in characterizing time-dependent corrosion of AZ91 and AM50 magnesium alloys characterization with respect to their microstructures. Journal of The Electrochemical Society. 2001;**148**:B489-B496. DOI: 10.1149/1.1415722

[29] Hassel AW, Lohrengel MM. The scanning droplet cell and its application to structured nanometer oxide films on aluminium. Electrochimica Acta. 1997;**42**:3327-3333. DOI: 10.1016/S0013-4686(97)00184-9

[30] Kollender JP, Voith M, Schneiderbauer S, Mardare AI, Hassel AW. Highly customisable scanning droplet cell microscopes using 3D-printing. Journal of Electroanalytical Chemistry. 2015;**740**:53-60. DOI: 10.1016/j.jelechem.2014.12.043

[31] Krawiec H, Stanek S, Vignal V, Lelito J, Suchy JS. The use of microcapillary techniques to study the corrosion resistance of AZ91 magnesium alloy at the microscale. Corrosion Science. 2011;**53**:3108-3113. DOI: 10.1016/j.corsci.2011.05.054

[32] Barranco V, Carmona N, Villegas MA, Galvan JC. Tailored sol-gel coatings as environmentally friendly pre-treatments for corrosion protection. ECS Transactions. 2010;**24**:277-290. DOI: 10.1149/1.3453622

[33] Mueller WD. Electrochemical techniques for assessment of corrosion behaviour of Mg and Mg-alloys. BioNanoMaterials. 2015;**16**:31-39. DOI: 10.1515/bnm-2015-0006

[34] Grundmeier G, Jüttner K, Stratmann M. Novel electrochemical techniques in corrosion research. In: Cahn RW, Haasen P, Kramer EJ, editors. Materials Science and Technology. Weinheim: Wiley-VCH Verlag GmbH & Co. KGaA; 2013. pp. 287-376. DOI: 10.1002/crat.2170300811

[35] Barranco V, Thiemann P, Yasuda HK, Stratmann M, Grundmeier G. Spectroscopic and electrochemical characterisation of thin cathodic plasma polymer films on iron. Applied Surface Science. 2004;**229**:87-96. DOI: 10.1016/j.apsusc.2004.01.046

[36] Thomas S, Izquierdo J, Birbilis N, Souto RM. Possibilities and limitations of scanning electrochemical microscopy of Mg and Mg alloys. Corrosion. 2015;**71**:171-183. DOI: 10.5006/1483

[37] Jamali SS, Moulton SE, Tallman DE, Forsyth M, Weber J, Wallace GG. Evaluating the corrosion behaviour of magnesium alloy in simulated biological fluid by using SECM

to detect hydrogen evolution. Electrochimica Acta. 2015;**152**:294-301. DOI: 10.1016/j.electacta.2014.11.012

[38] Liu W, Cao F, Xia Y, Chang L, Zhang J. Localized corrosion of magnesium alloys in NaCl solutions explored by scanning electrochemical microscopy in feedback mode. Electrochimica Acta. 2014;**132**:377-388. DOI: 10.1016/j.electacta.2014.04.044

[39] Izquierdo J, Nagy L, Bitter I, Souto RM, Nagy G. Potentiometric scanning electrochemical microscopy for the local characterization of the electrochemical behaviour of magnesium-based materials. Electrochimica Acta. 2013;**87**:283-293. DOI: 10.1016/j.electacta.2012.09.029

[40] Salleh SH, Thomas S, Yuwono JA, Venkatesan K, Birbilis N. Enhanced hydrogen evolution on $Mg(OH)_2$ covered Mg surfaces. Electrochimica Acta. 2015;**161**:144-152. DOI: 10.1016/j.electacta.2015.02.079

[41] Bastos AC, Quevedo MC, Karavai OV, Ferreira MGS. Review—On the application of the scanning vibrating electrode technique (SVET) to corrosion research. Journal of The Electrochemical Society. 2017;**164**:C973-C990. DOI: 10.1149/2.0431714jes

[42] Williams G, Neil McMurray H. Localized corrosion of magnesium in chloride-containing electrolyte studied by a scanning vibrating electrode technique. Journal of The Electrochemical Society. 2008;**155**:C340-C349. DOI: 10.1149/1.2918900

[43] Isaacs HS, Vyas B. Scanning reference electrode techniques in localized corrosion. In: Mansfeld F, Bertocci U, editors. STP727 Electrochemical Corrosion Testing, American Society for Testing and Materials. Philadelphia: ASTM; 1981. DOI: 10.1520/STP28024S

[44] Isaacs HS. Limitations of in situ current density mapping for vibrating electrodes close to metal surfaces. Corrosion. 1990;**46**:677-679. DOI: 10.5006/1.3585167

[45] Wilson BP, Searle JR, Yliniemi K, Worsley DA, McMurray HN. Effect of probe tip inclination on the response of the scanning vibrating electrode technique to an idealised pit-like feature. Electrochimica Acta. 2012;**66**:52-60. DOI: 10.1016/j.electacta.2012.01.062

[46] Williams G, McMurray HN, Grace R. Inhibition of magnesium localised corrosion in chloride containing electrolyte. Electrochimica Acta. 2010;**55**:7824-7833. DOI: 10.1016/j.electacta.2010.03.023

[47] Williams G, Birbilis N, McMurray HN. The source of hydrogen evolved from a magnesium anode. Electrochemistry Communications. 2013;**36**:1-5. DOI: 10.1016/j.elecom.2013.08.023

[48] Montemor MF, Simões AM, Carmezim MJ. Characterization of rare-earth conversion films formed on the AZ31 magnesium alloy and its relation with corrosion protection. Applied Surface Science. 2007;**253**:6922-6931. DOI: 10.1016/j.apsusc.2007.02.019

[49] Montemor MF, Ferreira MGS. Electrochemical study of modified bis-[triethoxysilylpropyl] tetrasulfide silane films applied on the AZ31 Mg alloy. Electrochimica Acta. 2007;**52**:7486-7495. DOI: 10.1016/j.electacta.2006.12.086

[50] Galio AF, Lamaka SV, Zheludkevich ML, Dick LFP, Müller IL, Ferreira MGS. Inhibitor-doped sol-gel coatings for corrosion protection of magnesium alloy AZ31. Surface and Coatings Technology. 2010;**204**:1479-1486. DOI: 10.1016/j.surfcoat.2009.09.067

[51] Simões AM, Battocchi D, Tallman DE, Bierwagen GP. SVET and SECM imaging of cathodic protection of aluminium by a Mg-rich coating. Corrosion Science. 2007;**49**:3838-3849. DOI: 10.1016/j.corsci.2007.03.045

[52] Bierwagen G, Battocchi D, Simões A, Stamness A, Tallman D. The use of multiple electrochemical techniques to characterize Mg-rich primers for Al alloys. Progress in Organic Coatings. 2007;**59**:172-178. DOI: 10.1016/j.porgcoat.2007.01.022

[53] Isaacs HS, Kendig MW. Determination of surface inhomogeneities using a scanning probe impedance technique. Corrosion. 1980;**36**:269-274. DOI: 10.5006/0010-9312

[54] Huang VM, Wu S-L, Orazem ME, Pébère N, Tribollet B, Vivier V. Local electrochemical impedance spectroscopy: A review and some recent developments. Electrochimica Acta. 2011;**56**:8048-8057. DOI: 10.1016/j.electacta.2011.03.018

[55] Bayet E, Huet F, Keddam M, Ogle K, Takenouti H. A novel way of measuring local electrochemical impedance using a single vibrating probe. Journal of The Electrochemical Society. 1997;**144**:L87-L90. DOI: 10.1149/1.1837562

[56] Baril G, Blanc C, Keddam M, Pébère N. Local electrochemical impedance spectroscopy applied to the corrosion behavior of an AZ91 magnesium alloy. Journal of The Electrochemical Society. 2003;**150**:B488-B493. DOI: 10.1149/1.1602080

[57] Wittmann MW, Leggat RB, Taylor SR. The detection and mapping of defects in organic coatings using local electrochemical impedance methods. Journal of The Electrochemical Society. 1999;**146**:4071-4075. DOI: 10.1149/1.1392593

[58] Baril G, Galicia G, Deslouis C, Pébère N, Tribollet B, Vivier V. An impedance investigation of the mechanism of pure magnesium corrosion in sodium sulfate solutions. Journal of The Electrochemical Society. 2007;**154**:C108-C113. DOI: 10.1149/1.2401056

[59] Barranco V, Carmona N, Galván JC, Grobelny M, Kwiatkowski L, Villegas MA. Electrochemical study of tailored sol-gel thin films as pre-treatment prior to organic coating for AZ91 magnesium alloy. Progress in Organic Coatings. 2010;**68**:347-355. DOI: 10.1016/j.porgcoat.2010.02.009

[60] Galvan JC, García-Galván FR, Barranco V, Jiménez-Morales A, Feliu S Jr. Waterborne sol-gel coatings doped with corrosion inhibitors for the protection of Mg-Al alloys. In: Proceedings of the Forty-Second Annual International Waterborne, High-Solids, and Powder Coatings Symposium; 8-13 February 2015; New Orleans. Raleigh: NC Lulu Press; 2015. pp. 349-365

Chapter 3

Corrosion Types of Magnesium Alloys

Rong-Chang Zeng, Zheng-Zheng Yin,
Xiao-Bo Chen and Dao-Kui Xu

Additional information is available at the end of the chapter

http://dx.doi.org/10.5772/intechopen.80083

Abstract

Magnesium (Mg) alloys are susceptible to corrosion in aggressive environments. Corrosion of Mg alloys depends greatly on their composition and microstructure (grain size, the size, shape and distribution of second phases), post-processing and media. In most cases, localized corrosion, such as pitting corrosion and filiform corrosion, generally occurs due to microgalvanic corrosion between the intermetallic compounds and their neighboring α-Mg matrix. However, open literature reported that several corrosion morphologies, that is, intergranular corrosion (IGC) and exfoliation corrosion (EFC), cannot appear on Mg alloys. In this chapter, all typical corrosion modes of Mg alloys and influencing factors are introduced, including general corrosion, galvanic corrosion, pitting corrosion, filiform corrosion, IGC, EFC, stress corrosion cracking (SCC), corrosion fatigue (CF) and so on. The focus is laid on pitting corrosion and EFC. Corrosion mechanisms of Mg alloys are also discussed.

Keywords: magnesium alloys, corrosion, intermetallic compounds, microstructure, biomaterials

1. Introduction

Magnesium (Mg) and its alloys can find extensive applications in transportation, 3C (computer, communication and consumer electronics) products and biomedical fields due to the lightweight, high specific strength and good biocompatibility [1, 2]. Mg alloys are susceptible to deterioration in an aggressive solution or harsh environment due to their low corrosion potential [3]. Like other metals, corrosion modes of Mg alloys can be divided into uniform or general corrosion and localized corrosion based on the phenomenon from electrochemical, composition and microstructural perspectives on macroscale. The latter predominately includes galvanic corrosion, pitting corrosion, filiform corrosion, intergranular corrosion (IGC), exfoliation corrosion (EFC), crevice corrosion, stress corrosion cracking (SCC), corrosion fatigue (CF) and erosion corrosion. EFC and crevice corrosion, however, have been rarely reported.

Corrosion form is fundamentally influenced by metallurgical factors such as chemical composition, grain size and shapes, size, shape and distribution of secondary phases or intermetallic compound particles, inclusions, solute-aggregated grain boundaries (GBs), crystallographic orientations and dislocation density. Moreover, post-processing (i.e., extrusion and rolling) and post heat-treatment (i.e., T4, T5 and T6) exert a significant impact on corrosion mechanism. These treatments can result in evident changes in microstructure and stress. Furthermore, corrosion mode is also highly related to chlorine ion content and pH value of the corrosive solution. Localized corrosion is prone to occur under low pH value or acidic and neutral solution.

In addition, modern characterization techniques such as environmental scanning electron microscopy (ESEM), scanning electrochemical microscopy (SECM) [4] and Kelvin probe force microscopy (KPFM) [5] open a door for understanding corrosion mechanisms of Mg alloys at a microscale level.

Investigation with ESEM shows that pitting corrosion of Mg alloy AZ91 merely emerges at the sites (AlMn phases and scratches) with segregation of salt particles [6]. Application of SECM reveals that pitting corrosion of Mg is initiated from film-free region [4]. The concentration of Cl^- ions and composition of the surface film dominate the incubation and growth of active sites in film-free region. Higher concentrated chloride ions result in rapid initiation and propagation of active spots.

It is critical for design engineers to have an insight into the corrosion mechanisms and morphology of Mg alloys and to avoid sudden pre-failure of Mg parts.

2. Forms of corrosion

2.1. Uniform corrosion

Uniform corrosion can be observed on solution (T4)-treated and peak-aged WE43 alloy in 3.5 wt.% NaCl solution saturated with $Mg(OH)_2$ [7]. This scenario for the T4-treated Mg alloy relates to the formation of a protective corrosion product layer, which is composed of an inner MgO layer and an outer $Mg(OH)_2$ layer. The improvement in corrosion resistance for the aging treated alloy is ascribed to the interactions between the finely distributed precipitates and the growth of corrosion reactions.

2.2. Galvanic corrosion

Galvanic corrosion is a common corrosion feature of Mg alloys. For example, galvanic corrosion may occur when Mg parts are connected or welded with the steels in lightweight powertrains and automotive bodies. In human body fluids, severe galvanic corrosion emerges when degradable Mg screws are connected with the bone plates made of titanium and stainless steel. This scenario is attributed to Mg that is located in the most active end in galvanic series of metals. Therefore, Mg and its alloys become the anode when connected to cathodic metals such as stainless steels and titanium alloys. This kind of galvanic corrosion is also termed bimetallic corrosion.

In some cases, galvanic corrosion occurs between two layers with different grain size in the same alloy. For instance, hot extruded Mg alloy ZK60 shows coarse grains in the outer layer and fine grains in the interior or middle layer [8]. The outer layer with a higher open corrosion potential (OCP) is passivated, while the middle layer with a lower OCP is preferentially corroded. It is the initial OCP rather than grain size that decides the initiation of galvanic corrosion [8].

At a micro-scale level, micro-galvanic corrosion may occur between the α-Mg matrix with a lower OCP and its surrounding second phases or intermetallic compounds with a higher OCP in Mg alloys. The matrix is, thus, anodic and preferentially corroded. This theory is used to explain the mechanism of localized corrosion, e.g., pitting corrosion [9], IGC [8] and EFC [5].

The nobler metals have a direct influence on the corrosion degree of the connected Mg alloys. The potential difference between Mg and other metals will be different, which can generate disparate galvanic couples and galvanic current. The galvanic current increases with the potential difference, that is, the bigger potential difference between Mg and dissimilar metal, the easier for Mg subject to corrode. The intensity of corrosion depends on the relative Volta potential differences of the micro-constituents, as well as their amounts, compositions and distributions within the alloys by KPFM [10].

The variations of pH may change the electrode reaction and polarity of galvanic metals. For instance, in an Al-Mg galvanic coupling in a neutral or weakly alkaline NaCl solution, Al acts as cathode. But with the dissolution of Mg, the solution changes into alkaline and aluminum turns into anode, which is the alkalinized effect of corrosion products of Mg alloys. Therefore, in an alkaline medium, that is, pH > 8.5, Mg alloys are basically non-corrosive. Mg alloys in acidic and neutral solutions, except for chromic acid, phosphoric acid and hydrofluoric acid that can form passive film, are susceptible to corrosion. When pH > 10.5, due to the formation of a protective $Mg(OH)_2$ film, Mg alloys have certain corrosion resistance. However, if a solution contains Cl^-, corrosion rate of Mg alloys will increase greatly.

Geometry factors such as cathode/anode area ratio, insulation distance between cathode and anode, depth of the dissolved film covering the galvanic couples and interactive mode of interaction caused by galvanic coupling can also affect the corrosion of Mg alloys.

The distance between galvanic couples is also important for corrosion behavior of Mg alloys. The distance of the galvanic current of the corrosion behavior of Mg alloys is limited. Corrosion generally occurs near the edge of such a specimen [11]. The declining amplitude decreases with the increase in the distance similarly to an exponential decay.

2.3. Pitting corrosion

Pitting corrosion is one of the most common corrosion types of Mg alloy. The generation of pits is related to chemical composition and specific microstructural features at the exposed Mg alloy surface. Chemical composition, grain size, heat treatment condition, morphology and distribution of the second phases all have an important influence on the corrosion of Mg alloys.

For sand-cast Mg alloy ZE41-T5 immersed in aggressive environment, pits initiate in the α-Mg grains neighboring T-phase (Mg_7Zn_3RE) at grain boundaries (GBs) [12]. This is attributed to galvanic corrosion, which is driven by the potential difference between the T-phase and α-Mg phase. In a subsequent immersion, the Zr-rich zones located in the center of grains become preferential corrosion sites. Small pits appear in the zones and severe attack develops.

In most cases, second phases that is, $Mg_{17}Al_{12}$ and Al_8Mn_5 are cathodic relative to α-Mg matrix [1], whereas in very limited cases, the second phases, containing rare earth (RE) and Mg in Mg alloy GW93 (Mg-Y-Gd), are more active than Mg matrix [13].

The microstructure constituents of Mg alloys are usually characterized by the α-Mg matrix and second phase or intermetallic phases, for instance, β ($Mg_{17}Al_{12}$), $MgSi_2$, AlMn(Fe), MgZn, Mg_2Zn [14], Mg_2Ca, Mg_2Sn [15], Mg_2Cu, Al_8Mn_5, I-phase (Mg_3Zn_6Y) [16] and T-phase [12]. Because in most cases, the electrode potential of these second phases is much higher than that of the α-Mg substrate, dissolution tends to emerge in the α-Mg matrix around second-phase particles. The presence of small Mg_2Ca particles coexisted with Fe and Si results in pitting corrosion of Mg-Ca alloys [17].

Occasionally, pitting corrosion is competitive when a number of second phases exist in Mg alloys. In extruded Mg alloy AZ80, there may be three kinds of intermetallic compounds: $Mg_{17}Al_{12}$, Al_8Mn_5 and Mg_2Si. Voltaic potential differences between them and the α-Mg matrix phase or contact potential difference are not the same. And corrosion preferentially occurs around AlMnFe second phase due to the fact that the potential difference between the AlMnFe phase and matrix is the biggest [18].

For Mg-Al alloys, pitting corrosion occurs along the net structure of $Mg_{17}Al_{12}$ selectively. TGC will take place in die casting Mg alloys AM50 and AM20 caused by coring. At first, corrosion pits were formed in Mn-bearing particles, then developed into corrosion cracks and extended and terminated in the high Al region or GBs of the crystals.

Pitting corrosion can be affected by grain size. Refined grains make the corrosion morphology from pitting to homogeneous or uniform corrosion.

The microstructure of an extruded Mg alloy AM60 is composed of α-Mg matrix, β phase along GBs and granular AlMn phase in grains [19]. The Al/Mn atomic ratio in AlMn phase is 1.12–1.15:1. In addition, there are a small number of Mg, Si elements and a small amount of deformation twins. The GB mainly consists of β phase. The AlMn phase is major in the grain that forms a cubic shape. There are both fine and coarse grains in the microstructure. Simultaneously, recrystallized grains form in the coarse grains. This result is due to the hot extrusion processing. The thick plate has a faster cooling rate for the outer layer and relatively slower cooling rate in the interiors, so localized recrystallization and grain abnormal growth phenomenon may occur.

In addition, pH value has a great influence on corrosion morphology of Mg alloys [19]. In acidic and neutral NaCl solutions, the α-Mg matrix phase of Mg alloy AM60 around AlMn particles is preferential to corrode. Thus, pitting corrosion of Mg alloy AM60 occurs. This may be the result of local galvanic corrosion due to a higher voltage of the AlMn phase than α-Mg matrix and β phase. In alkaline solutions, uniform corrosion of high alumina regions is observed with cellular corrosion morphology. At pH 12, the sample surface has only a few shallow corrosion pits, which is well protected. From the morphology of the pits, the site of incipient pitting corrosion is related to the AlMn phase. The pits form on α-Mg in the area adjacent to the AlMn phase and there are many cracks in the matrix near the pits [19]. The β phase does not play a role in the corrosion process of the extruded AM60 alloy.

Figure 1 illustrates the pitting corrosion mechanism of extruded AM60 alloy. The oxide film formed on Mg alloy surface is discontinuous, which provides the adsorption sites for anions such as Cl^-. Here, the corrosion pits of Mg alloys are hemispherical. Cl^- ions are adsorbed in α phase adjacent to the AlMn particles. At the self-corrosion potential (−1.53 V/SCE), the passivation film is broken, the α phase begins to dissolve and then pitting corrosion initiates. The AlMn particle in the pits is the cathode phase, while the α-Mg phase is the anode. There are a large number of Mg^{2+} and OH^- ions in corrosion pits. When Mg^{2+} ions reach the maximum solubility, $Mg(OH)_2$ will be formed and pH value of the solution will stabilize around 10.5. Hereby, saturated Mg^{2+} ion leads to precipitation of $Mg(OH)_2$ at the bottom of the pit. And partial Mg^{2+} and OH^- ions may diffuse out of the pit with H_2 evolution, leading to the formation of $Mg(OH)_2$ precipitate outside the rim of the corrosion pit. This is the so-called occluded cell.

When an extruded Mg-1.21Li-1.12Ca-1Y alloy was immersed in Hank's solution for the first 15 min, pitting corrosion occurred at the interface of Mg_2Ca/α-Mg instead of $Mg_{24}Y_5$/α-Mg [9]. Although the potential of the Mg_2Ca particle is higher than that of the α-Mg matrix phase, the dissolution rate of Mg_2Ca particles is higher than that of α-Mg matrix phase. After subsequent immersion of 1 h, corrosion product film formed around the pitting corrosion and filiform corrosion occurred. Pitting initiation in Mg_2Ca phase α-mg matrix forms corrosion products such as $MgCO_3$, $CaCO_3$, $LiCO_3$, $Mg_3(PO_4)_2$, $Ca_3(PO_4)_2$ and Li_3PO_4 that can make the corrosion product layer denser. During the electrochemical reaction, the surface of Mg_2Ca releases OH^- and H_2 and Mg^{2+} is produced by the dissolution of α-Mg. As the reaction proceeds, $Mg(OH)_2$ precipitates on the surface of Mg_2Ca phase. Mg^{2+} reacts further with HCO_3^- or CO_3^{2-} to form $MgCO_3$ precipitates [9].

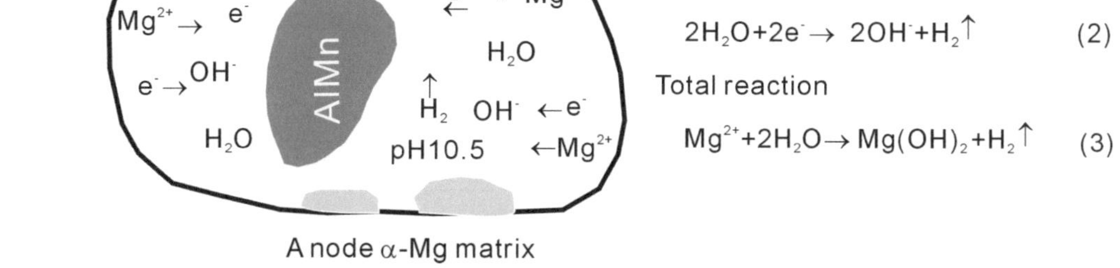

Figure 1. Model of pitting corrosion for Mg alloy AM60 [3].

The pitting corrosion occurs on as-cast dual-phase Mg-Li-Ca alloy due to the presence of Mg_2Ca particles, whereas uniform corrosion appears after an extrusion processing [20]. For the extruded Mg-Li-Ca alloy with refined microstructure, the tubular paths in the oxide film were jammed and sealed by the formation of an additional amorphous hydrated layer (**Figure 2**) and the chemical compounds, that is, LiOH, $Mg(OH)_2$, $CaCO_3$ and $MgCO_3$, the PBR values of which are between 1 and 2. In this scenario, the formation rate of the corrosion products LiOH, $Mg(OH)_2$, $CaCO_3$ $MgCO_3$ and $Ca_xMg_y(PO_4)_z$ is greater than that of the degradation rate of the Mg-Li-Ca alloy, and hence the dissolution of the alloy may be readily self-inhibited.

Pitting corrosion of Mg-Li alloys can be suppressed through grain refinement and formation of quasicrystalline or icosahedral phase (I-phase). I-phase eutectic pocket at GBs can suppress pitting corrosion and filiform corrosion of Mg-Li alloy with the presence of Zn and Y, regardless of Mg-6% Li-6% Zn-1.2% Y alloy containing a higher volume fraction of β-Li phase [16, 21].

2.4. Filiform corrosion

Filiform corrosion is also the most common type of corrosion for Mg alloys. Filiform corrosion of steel surfaces often occurs under coatings in high humidity environments. The filiform corrosion of Mg alloys is as common as pitting corrosion, but different from filiform corrosion of steels. Filiform corrosion is common on bare Mg alloy surfaces exposed to air and NaCl solution.

The pitting corrosion and filiform corrosion of Mg alloys usually occur simultaneously. Filiform corrosion originates from corrosion pits and extends forward along the active area. Corrosion of Mg alloys mainly is hydrogen evolution corrosion rather than oxygen absorption corrosion or oxygen concentration corrosion. Filiform corrosion is caused by active corrosive cells moving through the metal surface. Filiform corrosion of Mg is driven by the difference in oxygen concentration between filament head and filament tail, and thus proposed a filiform corrosion model for Mg [22]. Earlier literature reported that filiform corrosion occurs under protective coatings and anodic oxide layers, and pure Mg without coatings does not undergo filiform corrosion [1, 23].

Chloride-induced fliform corrosion of organic-coated Mg is studied by Williams *et al.* [24]. The rate of filiform corrosion advance is shown to be insensitive to the presence of oxygen, but highly dependent upon the relative humidity of the holding environment. The filiform corrosion of Mg-Li alloy produces high concentration of OH^- reacting with Mg^{2+} and Li^+ to form $Mg(OH)_2$ and LiOH, which makes filament tail alkalized, passivates and covers the corrosion products of $Mg(OH)_2$ [25]. The hydrolysis reaction acidifies the corrosion pits and becomes a filament tip. Then the tip continues to propagate forward.

2.5. Intergranular corrosion (IGC)

An earlier review [26] pointed out that IGC does not occur on Mg and its alloys. It is regarded that the attack does not enter the grain along GBs, due to the fact that GBs are almost always cathodic relative to the grain interiors. Therefore, grains are anodic with regard to GBs, and the area next to GBs will be corroded. But this is not regarded as IGC, but granular corrosion, because corrosion tends to concentrate in areas adjacent to GBs.

http://dx.doi.org/10.5772/intechopen.80083

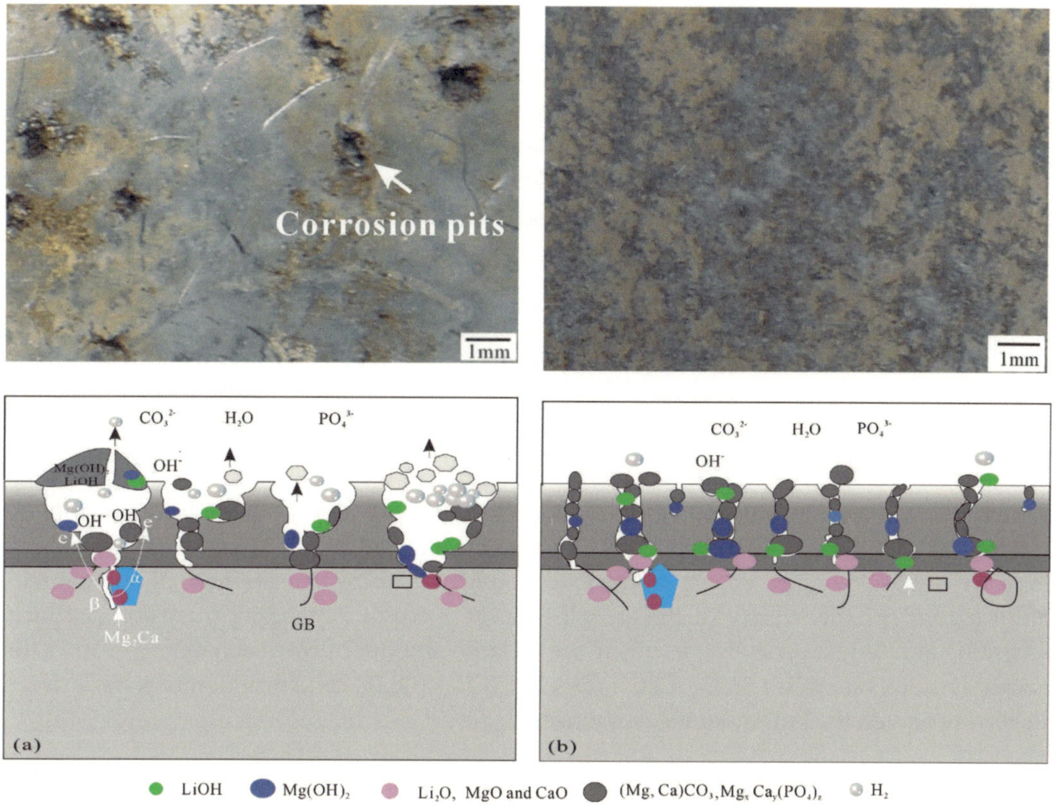

Figure 2. Schematic diagrams of the corrosion mechanisms of Mg-9Li-1Ca alloys: (a) pitting corrosion on the cast alloy and (b) tubular paths in the oxide film on the extruded alloy were jammed by corrosion products [20].

However, intergranular stress corrosion cracking (ISCC) has been observed in the process of stress corrosion of Mg alloys, which indicates that IGC of Mg alloys is truly possible. It has been found that the morphology of IGC exists in as-extruded Mg alloy AZ80 after T5 aging treatment [27]. The depth of this corrosion profile is shallow. The corrosion morphology is also due to the formation of micro-galvanic couple corrosion between $Mg_{17}Al_{12}$ and its surrounding matrix α-Mg, which is preferentially corroded [27].

From a microstructure perspective of the extrude Mg alloy ZK60, the fine second phase (Mg_2Zn) is dispersed along the GBs [8]. Thus, pearl-like corrosion pits have been observed along the GBs in the extrusion direction. For zirconium-containing aluminum-free Mg alloy ZK60, the second-phase Mg_2Zn on GBs is nobler relative to the α-Mg matrix phase, and corrosion occurs on the α-Mg matrix phase. Because zirconium is mainly distributed in the center of grains, the crystal is relatively stable. Further study discloses that the attack mainly occurs on the periphery of the grains, and no attack is seen to penetrate into the alloy along GBs. Therefore, the α-Mg matrix phase in the vicinity of GBs preferentially dissolves, which eventually leads to the whole grain peeling off. The cause of IGC is the formation of micro-galvanic corrosion between high potential finely distributed Mg_2Zn particles along the GBs and low potential matrix phase (α-Mg), which is typical IGC in Mg alloys for the first time.

2.6. Exfoliation corrosion (EFC)

EFC is usually being thought of as a special IGC. EFC on wrought Al alloys has been extensively reported [28–30]. Two essential factors exist in developing EFC: elongated grains in microstructure and different potential between the matrix and intermetallic compound precipitates at GBs [31]. However, EFC of Mg alloys has sparsely been reported. The EFC of cold-rolled Mg-14Li-1Al alloy appeared on both longitudinal and transverse surfaces [32]. At the initial stage of corrosion, oxide film formed rapidly on the surface of magnesium alloy. Then the oxide film was attacked by chlorine ions, leading to the dissolution of Mg in local areas. There corrosion micro-cracks obviously emerge along GBs, which eventually results in denudation. After annealing, the corresponding EFC disappeared. The introduction of Al suppresses the exfoliation behavior of Mg-14Li alloy. The excessive addition of Al leads to a decrease in corrosion resistance. The optimal corrosion resistance could be achieved by adding 3 wt.% Al element [33].

Recently, the corrosion morphologies of an as-extruded Mg-1Li-1Ca alloy have been investigated after immersion in 3.5 wt.% NaCl aqueous solution for 90, 120 and 150 days [5]. It is clearly that severe EFC occurs on the extruded Mg-1Li-1Ca plates (**Figure 3**) and exhibits a "lifting-up" of the surface of the samples by the force of expanding corrosion products emerging at GBs just beneath the subsurface. A large amount of corrosion debris, mainly the carbonates, for example, Li_2CO_3, $CaCO_3$ and $Mg_5(CO_3)_4(OH)_2$, can be evidently seen between the layers and on the bottom of the container.

The optical images in **Figure 4** disclose the bent outer layer and crack initiation sources and propagation directions. The cracks initiate along GBs or inside the grains, then coalescence, and arrest and propagation as well. The formation of voids or pores may be due to hydrogen blister and wedge effect resulted from the increase in volume of corrosion products.

Figure 5a demonstrates that open circuit potentials (OCPs) of the outer layer are much higher than those of the inner layer of as-extruded Mg-1Li-1Ca alloy. The gap in initial OCPs between them is greater than 88 mV/SCE. There is a risk of an increased corrosion rate for the less noble metal after galvanic coupling when the OCP difference is greater than 50 mV/SCE [8]. Macro-galvanic corrosion may, thus, occur between the inner and outer layers. **Figure 4b** shows the polarization curves of the inner zone and outer layer of the extruded Mg-1Li-1Ca alloy in 3.5 wt.% NaCl solution. Values of corrosion potential (E_{corr}) and corrosion current density (i_{corr}) were estimated through Tafel extrapolation fitting. E_{corr} values of the inner zone and outer layer are −1.639 ± 0.003 V/SCE and − 1.553 ± 0.009 V/SCE, respectively. Potential difference, $\triangle E_{corr}$ ($=E_{corr,\ outer\ layer} - E_{corr,\ inner\ zone}$ = 86 mV/SCE) of two layers is greater than 50 mV/SCE, implying the presence of galvanic corrosion.

KPFM was further used to evaluate Volta potential of the outer or skin layer and interior layer, as shown in **Figure 6**. For the interior zone, the Volta potential of extruded Mg-1Li-1Ca alloy fluctuates over a narrow range, while the skin layer in purple color corresponds to the most negative Volta potential. The Volta potential difference implies the possible presence of macro-galvanic corrosion between the interior zone and skin layer. This result is consistent with that of electrochemical tests (**Figure 5**).

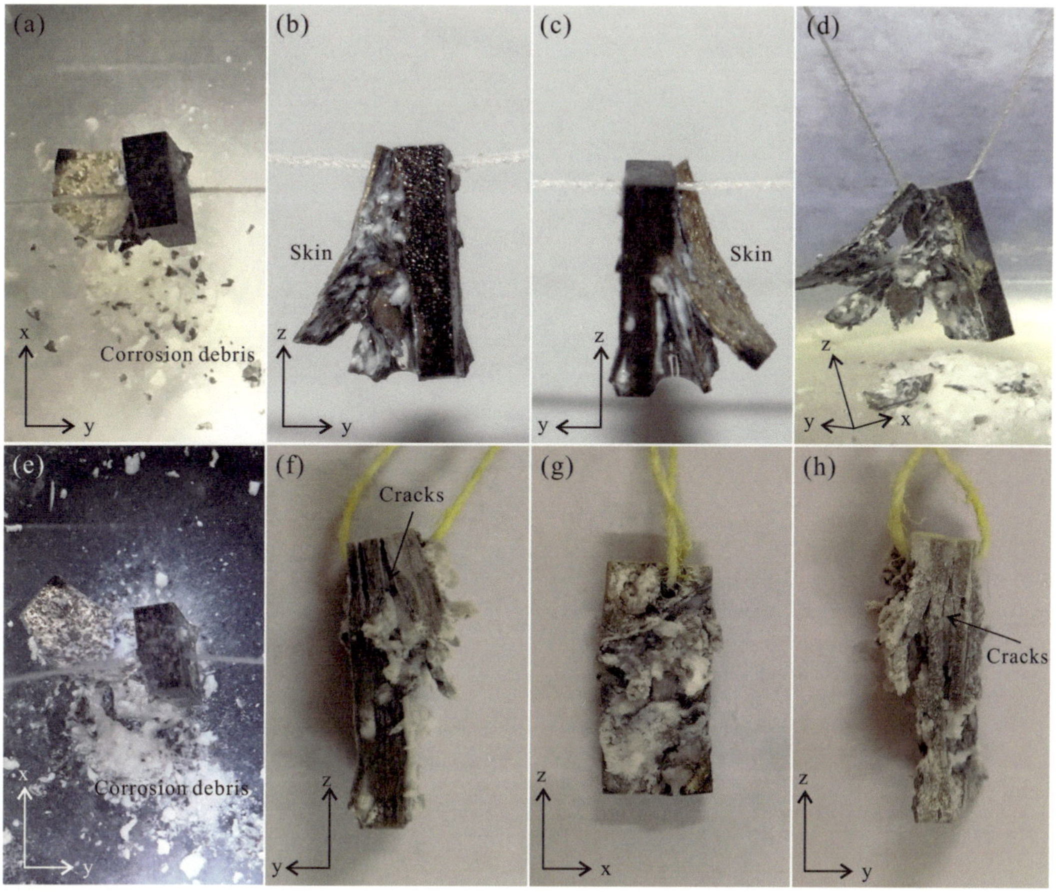

Figure 3. Corrosion macro-morphology of extruded Mg-1Li-1Ca specimens immersed in 3.5 wt.% NaCl aqueous solution after 90 days: (a)–(c); 120 days: (d) and (e); 150 days: (f)–(h) [5].

The corrosion current density, i_{corr}, for the inner layer and the outer layer of the extruded Mg-1Li-1Ca alloy is 5.789 ± 0.598 × 10^{-5} A/cm^2 and 4.718 ± 0.259 × 10^{-5} A/cm^2, respectively. Microstructure of the alloy is characterized by a fibrous structure with elongated α-Mg grains and fine intermetallic compound Mg_2Ca. The skin layer with more refined grains shows better corrosion resistance, whereas the interior with coarse grains and Mg_2Ca particles distribute along GBs and extrusion direction in a line. The inner zone exhibits a negative OCP in comparison with the skin layer. When the inner zone and outer layer are coupled together, the inner zone acts as anode and is preferentially dissolved, whereas the skin layer plays a cathodic role and is protected.

Accordingly, several factors such as microstructure (i.e., elongated microstructures and second phases) and stress dominate EFC of extruded Mg-1Li-1Ca alloy. EFC mechanisms of the extruded Mg-Li-Ca alloy are schematically illustrated in **Figure 7**. Exfoliation relates to both galvanic and wedge effects. On one hand, galvanic effect (**Figure 7a** and **b**) is resulted from the α-Mg grain and its neighboring second-phase Mg_2Ca particles, which are positioned

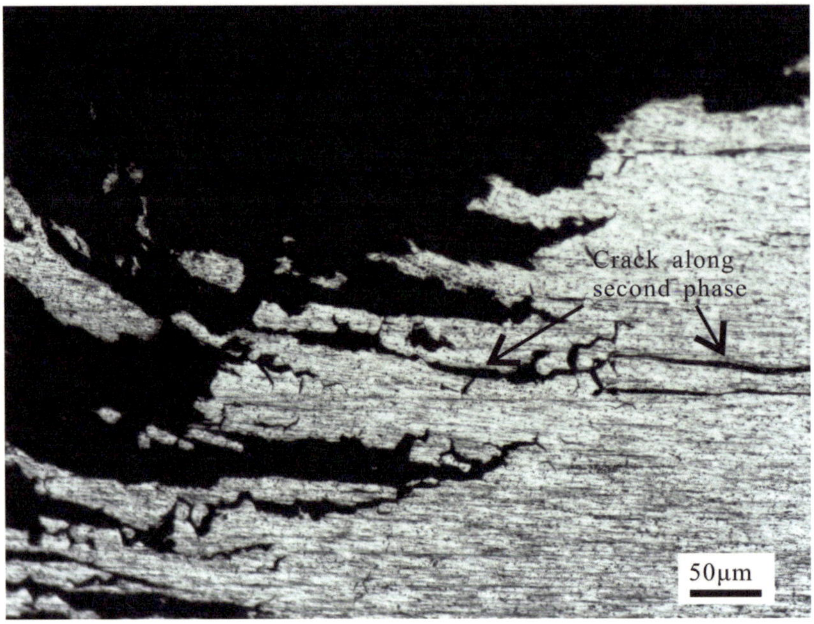

Figure 4. Optical image of cross-sectional morphologies of extruded Mg-1Li-1Ca alloy after an immersion in 3.5% NaCl aqueous solution for 21 days.

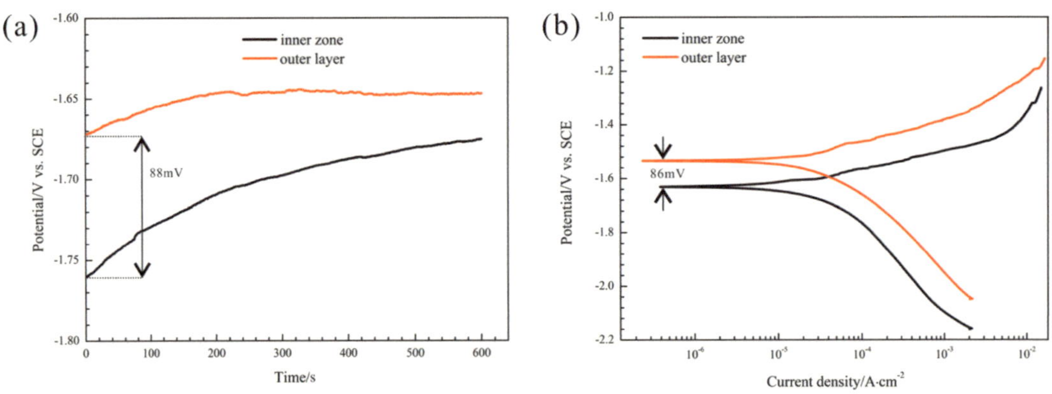

Figure 5. (a) Open circuit potential as a function of immersion time and (b) polarization curves of inner zone and outer layer of extruded Mg-1Li-1Ca alloy in 3.5 wt.% NaCl solution [5].

along the extrusion direction at both GBs and the interiors. On the other hand, the wedge effect (**Figure 7c**) is caused by the enhancement in volume of corrosion products, consisting of $Mg(OH)_2$, $Ca(OH)_2$ and carbonates such as Li_2CO_3, $CaCO_3$ and $Mg_5(CO_3)_4(OH)_2$ piled up in the cracks. Exfoliation initiates at corrosion pits and propagates along corrosion filaments (**Figure 7d**), in which filiform corrosion forms. Namely, EFC of the Mg-1Li-1Ca alloy emerged after pitting and filiform corrosion and was prone to stress corrosion. EFC of extruded Mg-Li-Ca alloys may be ascribed to a combining effect of pitting corrosion, filiform corrosion, IGC and SCC.

http://dx.doi.org/10.5772/intechopen.80083

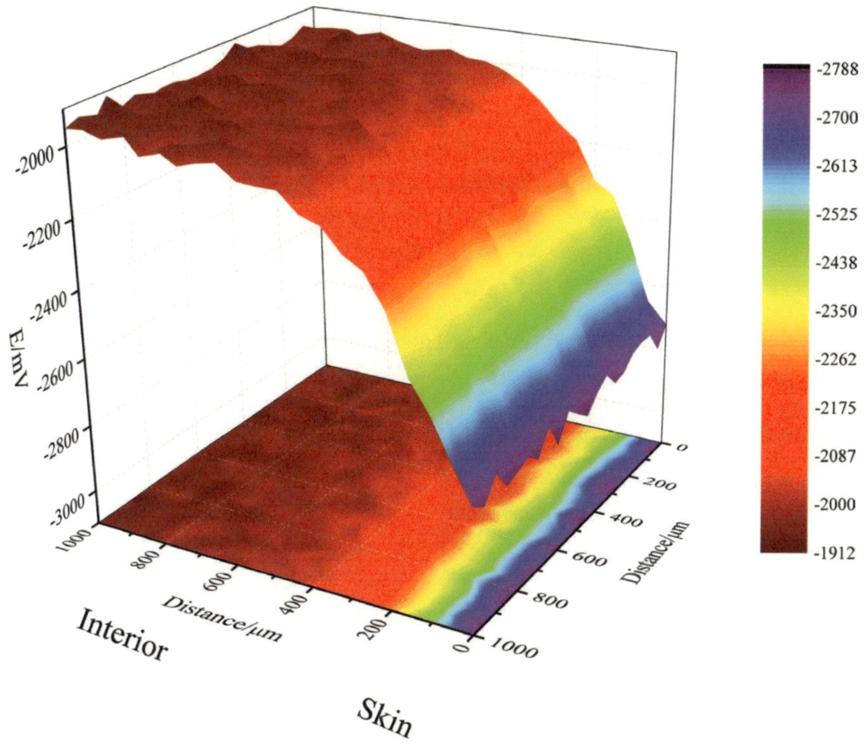

Figure 6. Surface Volta potential maps: longitudinal section of extruded Mg-1Li-1Ca alloy, detected by KPFM [5].

2.7. Crevice corrosion

Occurrence of crevice corrosion of metals, such as Fe, stainless steel and Al alloys, is generally considered to be driven by oxygen concentration (oxygen reduction cathode reaction), while for Mg alloys, the change in oxygen concentration can be neglected. Therefore, crevice corrosion of Mg alloys has been neglected before [34]. In 2011, crevice corrosion was firstly reported, which occurred in the crevice between specially treated pure Mg with unclear mechanism [35].

2.8. Stress corrosion cracking (SCC)

SCC of Mg alloys is the result of the combined action of electrochemistry and tensile mechanics. SCC can occur in any application when a stressed Mg-based component is subject to wet conditions. Basically, SCC mechanisms include anodic dissolution (AD) and hydrogen embrittlement (HE). The AD of Mg alloys is attributed to their high negative potential and the formation of a non-protective corrosion product film [36]. HE is demonstrated to widely take place due to the fact that the existence of H is inevitable. Catastrophic failure is expected when hydrogen is introduced into Mg alloys, which decreases the SCC initiation stress to below the operating stress [37]. So far, HE mechanism has not been clearly established. Pitting corrosion alleviated hydrogen into Mg alloy AZ80 and induced HE [36]. It is suggested that the cracking

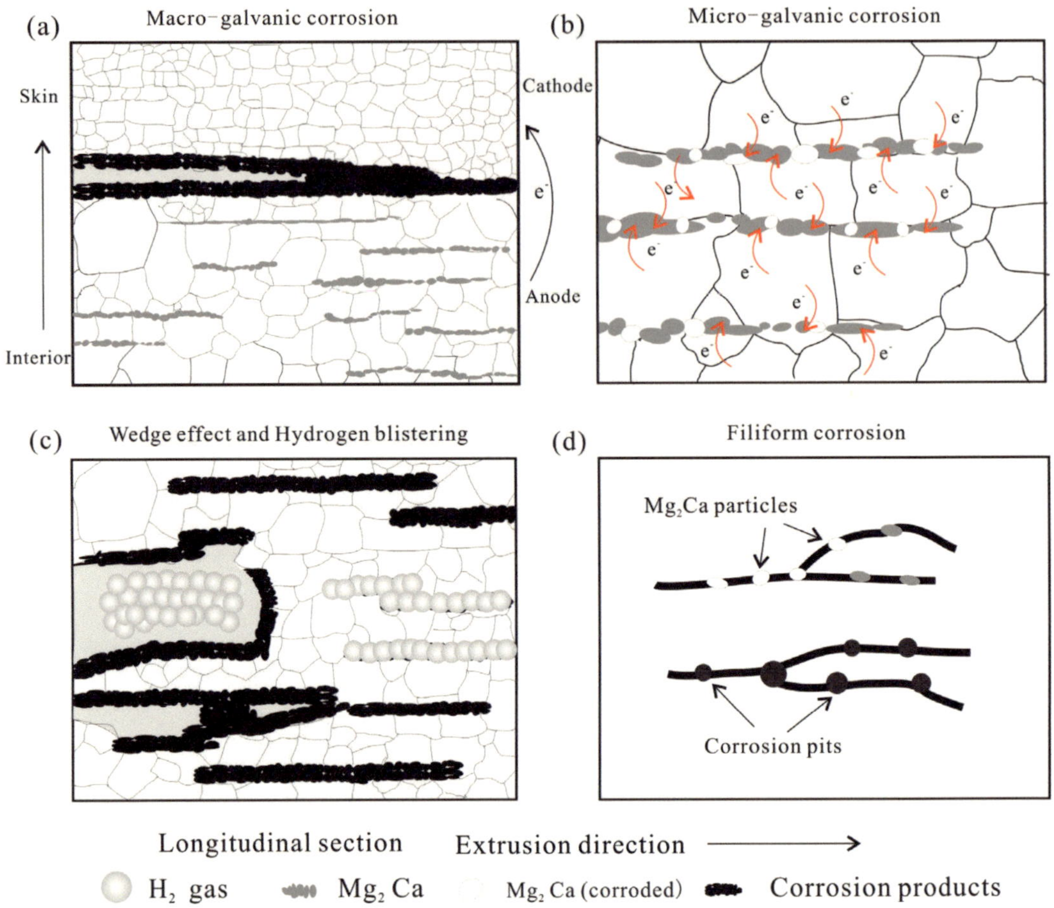

Figure 7. Schematic illustrations of EFC mechanism: (a) macro-galvanic corrosion between skin layer and interior zone, (b) micro-galvanic corrosion between Mg2Ca and α-Mg, (c) wedge effect and hydrogen blistering during EFC and (d) pitting corrosion and filiform corrosion [5].

mode of extruded Mg-7%Gd-5%Y-1%Nd-0.5%Zr (EW75) alloy is predominantly transgranular, but the partial intergranular stress corrosion cracking (IGSCC) can also be discerned at some localized area [38].

It is commonly recognized that transgranular SCC (TGSCC) of Mg alloys is a type of HE. HE models involve hydrogen-enhanced decohesion (HEDE), hydrogen-enhanced localized plasticity (HELP), adsorption-induced dislocation emission (AIDE) and delayed hydride cracking (DHC). The presence of hydride (MgH_2) phase on the fracture surface designates the DHC mechanism for a friction-stir welded magnesium alloy AZ31 [39].

One of the results of plastic deformation of polycrystalline materials with hexagonal structure is the formation of many deformation twins. Although the deformation twin plays a very important role in other fracture processes, the function on stress corrosion process is ignored. For example, at room temperature, the FCP of polycrystalline Mg alloy is related to twin [40] and GBs [41, 42]. Stress corrosion fracture surface of Mg-8.6Al alloy contains a lot of deformation twins [43]. Twin rotation and hydrides occur during slow tensile stress corrosion of extruded Mg alloy AZ31 [39]. The relationship between SCC and deformation twin crystal plane is unclear.

Alloying elements have different impacts on SCC [43]. Al is the most important element to produce stress corrosion sensitivity of Mg alloys. Mg-Al alloys have the highest stress corrosion sensitivity, and the sensitivity increases with Al content. An Al content of Mg alloys above 0.15–2.5 wt.% will result in SCC, with the maximum concentration at 6 wt.% Al. Zn also results in SCC sensitivity of Mg alloy. The addition of zirconium or RE elements (without Al) for Mg-Zn alloys such as ZK60 and ZE10 has moderate SCC resistance. Mg-Mn alloys are the most resistant to SCC due to the absence of Al and Zn.

Mg alloys also undergo SCC in distilled water [37]. Some anions can also accelerate stress corrosion of Mg alloys. SCC sensitivity of Mg alloys in 0.1 M neutral salt solution decreases in the following order: $Na_2SO_4 > NaNO_3 > Na_2CO_3 > NaCl > CH_3COONa$.

The SCC behaviors of Mg alloy ZK60 under different conditions, that is, thin electrolyte layer (TEL) and bulk solution, had been investigated [44]. Results indicated that Mg alloy ZK60 in the bulk solution exhibits a higher SCC susceptibility with a combined SCC mechanism of weaker AD and stronger HE compared to under TEL. They further revealed that cracking mode under both conditions was dominated by TGSCC feature and related to different HE mechanisms.

The effect of bovine serum albumin (BSA) to SCC of Mg alloy AZ91D in Hanks' solution was studied in [45]. The alloy was susceptible to SCC with and without the presence of BSA. In the presence of BSA, the alloy suffered greater SCC on account of greater susceptibility to pit formation.

2.9. Corrosion fatigue (CF)

CF is a key factor determining the service life of structural metals under cyclic mechanical loads [46]. For example, car wheels must be resistant to CF. Air seems to be an aggressive atmosphere for Mg alloys. Microstructure and loading frequency have a critical influence on fatigue life and fatigue crack propagation rate (FCPR) of Mg alloys in ambient atmosphere [47–49]. Since corrosion is time-dependent, CF of Mg alloys are sensitive to loading frequency. Under 10 Hz in air, fatigue life for as-extruded Mg alloy AM60 has a certain relationship with frequency [47]. But when frequency is above 10 Hz, fatigue life is independent of frequency. This may be due to the slip or twinning of plastic deformation of Mg alloys at low frequencies and the formation of the intrusions or extrusions on the sample surface. This time seems to be enough to form an oxide film, which leads to an irreversible plastic deformation in a fatigue process. Because the plastic deformation speed is higher at a higher frequency, there's no sufficient time for extrusion of Mg alloys to react with oxygen in air to form an oxide film. Therefore, the formation of oxide film in a process of plastic deformation can be avoided at high frequency.

An investigation demonstrated the coalescence of micro-void FCP mechanism of the extruded AZ80 at a constant load amplitude fatigue test by virtue of scanning electron microscopy (SEM), transmission electron microscopy (TEM) and Auger electron spectroscopy (AES) [50].

FCPR increases with a decreasing loading frequency and an increasing load ratio. AES results designate that fatigue can markedly facilitate the fracture oxidation of magnesium alloys. In addition, the occurrence of crack closure is ascribed to plasticity instead of oxide film.

The pH value and Cl^- ion-containing solutions can significantly affect the fatigue life of Mg alloys. Fatigue life of as-extruded Mg alloy AM60 in a neutral NaCl solution was the shortest and the longest in an alkaline NaCl solution [47]. CF strength of Mg alloys AZ91HP, AZ91-T6 and AM60 in 3.5% NaCl solution decreases significantly in comparison to that in air. In addition, Cl^-, Br^-, I^- and SO_4^{2-} ions accelerate the corrosion fatigue crack growth of Mg alloys [51, 52].

Relative humidity (RH) and temperature in air have an important effect on FCP for magnesium alloys [53]. Below 60 RH%, fatigue strength is no affected. At 90 RH% and higher, however, fatigue strength is significantly reduced. The effect of RH on the fatigue strength of extruded Mg alloy AZ61 has been investigated [54]. At 55 RH%, the fatigue limitation is 145–150 MPa between 20 and 50°C. At 80 RH% fatigue fracture occurs at the stress below the fatigue limitation, which means occurrence of CF. Corrosion pits are observed on the surface and fracture initiation area of the specimens. An increase in ambient temperature from 60 to 120°C results in a higher FCPR [53].

Heat treatment (i.e., artificial aging T5) gives rise to an improvement in volume fraction of β phase, tensile strength and hardness of Mg alloy AZ80, and the fatigue life in aggressive solution at lower stress level [55].

For HCP metals, the combination of twinning and fatigue deformation with the existing twins is the main form of fatigue deformation. Dislocations and slip bands stack along the mechanical twin bands. Twins hinder the movement of dislocations, cause dislocation pile-ups, improve the internal stress of the matrix, and result in the combination of dislocation and form fatigue cracks eventually. Basically, the preferential initiation of fatigue cracks for cast Mg alloys develops at stress concentration sites such as the casting defects and pores. Fatigue cracks for wrought Mg alloys preferentially initiate at inclusions, cyclic slip bands and twin boundaries [46].

In air, fatigue crack of Mg alloy AZ80 initiates at the inclusions in the surface and subsurface [55]. Corrosion pits are often the initiation sites for CF cracks. The formation of fatigue crack in an extruded Mg alloy AM60 relates to the AlMn particles. AlMn phase is the source of fatigue crack initiation in air. When it comes to a solution medium, the corrosion fatigue crack initiates at the corrosion pits around the AlMn phase particles [56]. AD is the CF mechanism of extruded Mg alloys.

CF behavior of die-cast and extruded Mg alloys can be different. The fatigue life of die-cast and especially extruded Mg alloys AZ91D, AM50 and AZ31 is significantly decreased in 3.5% NaCl solution [57]. Extruded alloys show a higher sensitivity to 3.5% NaCl solution compared to die-cast alloys. The CF life of extruded alloys is, however, longer than that of die-cast alloys.

Fatigue behavior of Mg alloys in simulated body fluids (SBF) is similar to that in 3.5 wt.% NaCl aqueous solution. Die-cast Mg alloy AZ91D showed a fatigue limit of 50 MPa at 10^7 cycles in air in comparison to 20 MPa at 10^6 cycles in SBF at 37°C [58]. The extruded WE43 alloy possesses a fatigue limit of 110 MPa at 10^7 cycles in air compared to 40 MPa at 10^7 cycles in SBF at 37°C. The fatigue cracks initiate at the micro-pores in air and at corrosion pits in SBF, respectively. Corrosion resistance of such two alloys decreases under cyclic loading compared to that in the static immersion test. A further study [59, 60] demonstrated that sand-cast Mg alloy AZ91D, high-purity Mg, Mg-Ca and Mg-Zn-Ca alloys are susceptible to CF in SBF. And the fatigue cracks initiate at the microstructural defects such as inclusions, cast pores, micro-cracks and corrosion pits. HE is responsible for the cracking of Mg alloy AZ91D [59].

The influence of bovine serum albumin (BSA) on CF of Mg alloys has recently been reported [61]. At stresses higher than the fatigue limit of Mg alloy AZ91D in air, mechanical factors play a predominant role. Hence, fatigue life is comparative for the alloys tested in air, Hank's solution in absence and presence of BSA. However, at lower stresses, Mg alloy AZ91D tested in Hank's solution + BSA discloses a longer fatigue life than that tested in a plain Hank's solution, indicating the inhibition impact of adsorbed BSA on corrosion and corrosion FCPR of the alloy.

Several strategies have been attempted to improve the resistance to CF. The shot-blasted Mg alloy AM60 exhibits an enhancement in fatigue strength in comparison to the die-cast one, which is ascribed to the fact that shot-blasting can induce hardening and compressive residual stress near surface region [62]. Nevertheless, coatings such as calcium phosphate and MgF_2 film can provide little improvement in CF resistance [63]. This is due to the degradation of the coating layer and existence of defects, which act as CF initiation sites under cyclical loading condition in SBF.

2.10. Erosion corrosion

Mg alloys have low surface hardness and poor resistance to wear. Wear resistance relates to load amplitude and loading frequency [64]. Wear loss of Mg alloys AZ91D and AM60B increases with normal loading force, but reduces with an increase in loading frequency. Corrosion and wear resistances can be greatly improved through micro-arc oxidation (MAO), a ceramic coating formed on the surface of Mg alloy AZ91 [65]. An $MgCO_3{\cdot}3H_2O$ film formed on Mg alloy AZ91 in HCO_3^- containing solutions is more compact and wear-resistant than porous $Mg(OH)_2$ film [66]. Further study reveals the wear diameter of Mg alloy AZ91 and its MAO coating in HCO_3^- containing solution is smaller than that in NaCl solution, indicating that the formation of corrosion products in HCO_3^- bearing solution can resist micro-abrasion to a certain extent. That is, HCO_3^- ions can promote the wear resistance of Mg alloy AZ91 with and without MAO coating. There is, however, a critical concentration of HCO_3^- for enhancement in wear resistance. During wear corrosion, oxide particles peel off from MAO coating, which is detrimental to the corrosion resistance of MAO-coated Mg alloy AZ91. Ni-P coating especially added with inorganic nano-oxide particles (i.e., TiO_2 and SiO_2) can effectively improve the wear resistance of pure Mg [40].

Flow-induced shear stress (FISS) accelerates general corrosion and localized corrosion (including pitting corrosion and erosion corrosion) due to the enhanced mass transfer and mechanical force [67]. FISS improves general corrosion rate, localized corrosion coverage ratios and depths and removal rate of corrosion products inside the corrosion pits compared to static conditions. Flow direction has a marked influence on corrosion resistance. More severe pitting corrosion and erosion corrosion were discerned on the back ends of the Mg-Zn-Ca alloy. And the corrosion product layer facing the flow direction was delaminated from Mg alloy AZ31 stent struts.

2.11. High temperature oxidation

Since Mg and its alloys are quite active and apt to react with surrounding oxidizing media such as oxygen, nitrogen and water vapor, oxidation-resistant and mechanical properties of Mg alloys are low at elevated temperatures. The oxidation resistance of metals depends mainly on the density and coverage of surface oxide films, which are also characterized by Pilling-Bedworth ratio (PBR). PBR is related to the relative volume of oxide and its metal. When PBR is less than one, oxide film of the metal is porous and loose. When PBR is equal to or greater than one, a complete oxide film forms on the surface of the metal. Thus, the metal can be protected by such an oxide film. Basically, PBR of Mg oxide film is below one, which indicates that the oxide films of Mg alloys cannot protect substrates from the attack from the environment.

Alloying with Ca, Be and Sn as well as RE elements can improve the oxidation-resistance of Mg alloys significantly. Studies on the oxidation properties of Mg-Y-Sn, Mg-Y and Mg-Sn alloys at 500°C show that Mg-Y-Sn and Mg-Y alloys demonstrate higher oxidation resistance compared with pure Mg and Mg alloy AZ31 [68]. The formation of a compact surface oxide film on Mg-Y-Sn and Mg-Y alloys after being oxidized at 500°C for 6 h can effectively protect the alloys from further oxidation. Corrosion resistance of oxidized Mg-Y-Sn and Mg-Y alloys can be enhanced in comparison with the initial non-oxidized Mg alloys. Our recent study [5] reveals that spark of micro-arc oxidation of Mg-Li-Ca alloys initiates at the α-Mg matrix with a lower content of Ca. This is due to the ignition inhibition of Ca.

There is, however, an issue on the oxide film of Ca-containing Mg alloys. Though the value of PBR for the oxides (MgO and CaO) of Mg-Ca alloys should be less than one, they exhibit good oxidation resistance. The working mechanism related to such an unexpected performance remains unclear.

2.12. Alternative corrosion modes

In many cases, several corrosion modes are observed on the same Mg alloys simultaneously or continuously. The corrosion process of Mg-Zn-Y-Zr alloy experiences three stages: galvanic corrosion, filiform corrosion and pitting corrosion in 0.1 M NaCl solution [69]. The corrosion mechanism is governed by the second phase containing Mg, Zn and Y, which is much nobler than α-Mg matrix. There is a huge potential difference between the intermetallic compounds and their α-Mg matrix. At the first stage, the intermetallic compounds create galvanic corrosion. At the mid-stage, the intermetallic compounds prevent propagation of filiform corrosion. And at the last stage, the intermetallic compounds accelerate pitting corrosion. Different from the potential difference of Mg-Zn-Y-Zr alloy, the filaments of filiform corrosion on Mg-8Li alloy can extend a long distance due to a low potential difference between β-Li phase and α-Mg phase.

Similarly, different types of corrosion emerge alternatively for Mg-Gd-Zn alloys [70]. Galvanic corrosion appears near second phase of $(Mg, Zn)_3Gd$ in the initial immersion, and filiform corrosion emerges after an immersion of 5 h. The difference of IGC developed in 16-h immersion is ascribed to the dissolution of $(Mg, Zn)_3Gd$ phase.

As mentioned earlier [5], after a long-term immersion, the extruded Mg-Li-Ca alloys experience synergistic influence by pitting corrosion, filiform corrosion and EFC as well as SCC.

Corrosion morphology changes with the constituents of the films and volume fraction of the second phases. For Mg-7Sn alloy, the corrosion mode and rate are concerned with the quantity of the second phase Mg_2Sn and tin content of the α-Mg matrix [15]. Pitting corrosion occurs when most of tin is in the presence of Mg_2Sn phase, while filiform corrosion emerges if most of tin is solutionized in the α-Mg matrix.

3. Conclusions

Various corrosion forms of Mg alloys are introduced in this chapter. The focus is on the novel findings on EFC of extruded Mg-Li-Ca alloy. The prestigious influencing factor may be the second phases, from which the initiation of almost all the electrochemical corrosion forms of Mg alloys originates. The intrinsic corrosion of Mg alloys may be attributable to micro-galvanic corrosion between the intermetallic compounds and their neighboring α-Mg matrix. A homogenous microstructure with fine grains and intermetallic compounds can suppress the tendency of localized corrosion and lead to general corrosion. The forms of corrosion of Mg alloys may vary with extending immersion time. Several modes of corrosion may vary alternatively on the same alloy. The second phases are not always positive and act as cathode relative to α-Mg matrix. The second phases with RE may be anodic relative to α-Mg matrix.

Acknowledgements

This work is supported by National Science Foundation of China (51571134) and SDUST Research Fund (2014TDJH104).

Conflict of interest

The authors declared that they have no conflicts of interest to this work.

Appendices and nomenclature

AIDE	Adsorption-induced dislocation emission
AD	Anodic dissolution

AES	Auger electron spectroscopy
BSA	Bovine serum albumin
CF	Corrosion fatigue
DHC	Delayed hydride cracking
ESEM	Environmental scanning electron microscope
EFC	Exfoliation corrosion
FCPR	Fatigue crack propagation rate
FISS	Flow-induced shear stress
GB	Grain boundary
HE	Hydrogen embrittlement
HEDE	Hydrogen-enhanced decohesion
HELP	Hydrogen-enhanced localized plasticity
IGC	Intergranular corrosion
ISCC	Intergranular stress corrosion cracking
KPFM	Kelvin probe force microscopy
Mg	Magnesium
MAO	Micro-arc oxidation
OCP	Open corrosion potential
PBR	Pilling-Bedworth ratio
RH	Relative humidity
SEM	Scanning electron microscopy
SCC	Stress corrosion cracking
SKP	Scanning Kalvin probe
SECM	Scanning electrochemical microscopy
SBFs	Simulated body fluids
TGSCC	Transgranular SCC
TEM	Transmission electron microscopy

Author details

Rong-Chang Zeng[1]*, Zheng-Zheng Yin[1], Xiao-Bo Chen[2] and Dao-Kui Xu[3]

*Address all correspondence to: rczeng@foxmail.com

1 School of Materials Science and Engineering, Shandong University of Science and Technology, Qingdao, China

2 School of Engineering, RMIT University, Carlton, Australia

3 CAS Key Laboratory of Nuclear Materials and Safety Assessment, Institute of Metal Research, Chinese Academy of Sciences, Shenyang, China

References

[1] Zeng RC, Jin Z, Huang WJ, Dietzel W, Kainer KU, Blawert C, et al. Review of studies on corrosion of magnesium alloys. Transactions of Nonferrous Metals Society of China. 2006;**16**:763-771. DOI: 10.1016/S1003-6326(06)60297-5

[2] Zeng RC, Dietzel W, Witte F, Hort N, Blawert C. Progress and challenge for magnesium alloys as biomaterials. Advanced Engineering Materials. 2010;**10**:B3-B14. DOI: 10.1002/adem.200800035

[3] Zeng RC, Cui LY, Ke W. Biomedical magnesium alloys: Composition, microstructure and corrosion. Acta Metallurgica Sinica. 2018;**54**. DOI: 10.11900/0412.1961.2018.00032

[4] Liu W, Cao F, Xia Y, Chang L, Zhang J. Localized corrosion of magnesium alloys in NaCl solutions explored by scanning electrochemical microscopy in feedback mode. Electrochimica Acta. 2014;**132**:377-388. DOI: 10.1016/j.electacta.2014.04.044

[5] Ding Z-Y, Cui L-Y, Zeng R-C, Zhao Y-B, Guan S-K, Xu D-K, et al. Exfoliation corrosion of extruded Mg-Li-Ca alloy. Journal of Materials Science & Technology. 2018;**34**:1550-1557. DOI:10.1016/j.jmst.2018.05.014

[6] Chen J, Wang J, Han E-H, Ke W. In situ observation of pit initiation of passivated AZ91 magnesium alloy. Corrosion Science. 2009;**51**:477-484. DOI: 10.1016/j.corsci.2008.11.024

[7] Chu P-W, Marquis EA. Linking the microstructure of a heat-treated WE43 Mg alloy with its corrosion behavior. Corrosion Science. 2015;**101**:94-104. DOI: 10.1016/j.corsci.2015.09.005

[8] Zeng R, Kainer KU, Blawert C, Dietzel W. Corrosion of an extruded magnesium alloy ZK60 component—The role of microstructural features. Journal of Alloys & Compounds. 2011;**509**:4462-4469. DOI: 10.1016/j.jallcom.2011.01.116

[9] Zeng RC, Qi WC, Zhang F, Cui HZ, Zheng YF. In vitro corrosion of Mg–1.21Li–1.12Ca–1Y alloy. Progress in Natural Science: Materials International. 2014;**24**:492-499. DOI: 10.1016/j.pnsc.2014.08.005

[10] Coy AE, Viejo F, Skeldon P, Thompson GE. Susceptibility of rare-earth-magnesium alloys to micro-galvanic corrosion. Corrosion Science. 2010;**52**:3896-3906. DOI: 10.1016/j.corsci.2010.08.006

[11] Jia J, Atrens A, Song G, Muster TH. Simulation of galvanic corrosion of magnesium coupled to a steel fastener in NaCl solution. Materials & Corrosion. 2015;**56**:468-474. DOI: 10.1002/maco.200403855

[12] Neil WC, Forsyth M, Howlett PC, Hutchinson CR, Hinton BRW. Corrosion of magnesium alloy ZE41—The role of microstructural features. Corrosion Science. 2009;**51**:387-394. DOI: 10.1016/j.corsci.2008.11.005

[13] Song YW, Shan DY, Han EH. Pitting corrosion of a rare earth Mg alloy GW93. Journal of Materials Science & Technology. 2017;**33**:954-960. DOI: 10.1016/j.jmst.2017.01.014

[14] Zeng RC, Wang L, Zhang DF, Cui HZ, Han EH. In vitro corrosion of Mg-6Zn-1Mn-4Sn-1.5Nd/0.5Y alloys. Frontiers of Materials Science. 2014;**8**:230-243. DOI: 10.1007/s11706-014-0256-6

[15] Liu XB, Shan DY, Song YW, Chen RS, Han EH. Influences of the quantity of Mg2Sn phase on the corrosion behavior of Mg–7Sn magnesium alloy. Electrochimica Acta. 2011;**56**:2582-2590. DOI: 10.1016/j.electacta.2010.12.030

[16] Xu DK, Han EH. Effect of quasicrystalline phase on improving the corrosion resistance of a duplex structured Mg–Li alloy. Scripta Materialia. 2014;**71**:21-24. DOI: 10.1016/j.scriptamat.2013.09.025

[17] Zeng R-C, Qi W-C, Cui H-Z, Zhang F, Li S-Q, Han E-H. In vitro corrosion of as-extruded Mg–Ca alloys—The influence of Ca concentration. Corrosion Science. 2015;**96**:23-31. DOI: 10.1016/j.corsci.2015.03.018

[18] Andreatta F, Apachitei I, Kodentsov AA, Dzwonczyk J, Duszczyk J. Volta potential of second phase particles in extruded AZ80 magnesium alloy. Electrochimica Acta. 2006;**51**:3551-3557. DOI: 10.1016/j.electacta.2005.10.010

[19] Zeng RC, Zhou WQ, Han EH, Ke W. Effect of pH values onas-extruded magnesium alloy AM60. Acta Metallurgica Sinica. 2005;**41**:307-311. DOI: 10.3321/j.issn:0412-1961.2005.03.017

[20] Zeng R-C, Sun L, Zheng Y-F, Cui H-Z, Han E-H. Corrosion and characterisation of dual phase Mg–Li–Ca alloy in Hank's solution: The influence of microstructural features. Corrosion Science. 2014;**79**:69-82. DOI: 10.1016/j.corsci.2013.10.028

[21] Xu DK, Han EH. Effects of icosahedral phase formation on the microstructure and mechanical improvement of Mg alloys: A review. Progress in Natural Science: Materials International. 2012;**22**:364-385. DOI: 10.1016/j.pnsc.2012.09.005

[22] Evans UR, Kruger J, Brown BF. Localized corrosion. In: Proceedings of the NACE International Corrosion Conference Series (NACE-3); 6-10 December 1971; Williamsburg, Virginia: NACE; 1974. p. 1215-1235

[23] Ghali E, editors. General, Galvanic, and Localized Corrosion of Aluminum and Its Alloys. Chichester: John Wiley & Sons, Inc.; 2010. 176-214 pp. DOI: 10.1002/9780470531778.ch5

[24] Williams G, Grace R. Chloride-induced filiform corrosion of organic-coated magnesium. Electrochimica Acta. 2011;**56**:1894-1903. DOI: 10.1016/j.electacta.2010.09.005

[25] Song YW, Shan DY, Chen RS, Han EH. Corrosion characterization of Mg–8Li alloy in NaCl solution. Corrosion Science. 2009;**51**:1087-1094. DOI: 10.1016/j.corsci.2009.03.011

[26] Song GL, Atrens A. Corrosion mechanisms of magnesium alloys. Advanced Engineering Materials. 1999;**1**:11-33. DOI: 10.1002/(SICI)1527-2648(199909)1:1<11::AID-ADEM11>3.0.CO;2-N

[27] Zeng R, Han E, Wei KE. Corrosion of artificial aged magnesium alloy AZ80 in 3.5 wt pct NaCl Solutions. Journal of Materials Science & Technology. 2007;**23**:353-358

[28] Reboul MC, Bouvaist J. Exfoliation corrosion mechanisms in the 7020 aluminium alloy. Materials & Corrosion. 1979;**30**:700-712. DOI: 10.1002/maco.19790301006

[29] Robinson MJ. Mathematical modelling of exfoliation corrosion in high strength aluminium alloys. Corrosion Science. 1982;**22**:775-790. DOI: 10.1016/0010-938X(82)90013-0

[30] Conde A, Damborenea JD. Electrochemical modelling of exfoliation corrosion behaviour of 8090 alloy. Electrochimica Acta. 1998;**43**:849-860. DOI: 10.1016/S0013-4686(97)00218-1

[31] Marlaud T, Malki B, Henon C, Deschamps A, Baroux B. Relationship between alloy composition, microstructure and exfoliation corrosion in Al–Zn–Mg–Cu alloys. Corrosion Science. 2011;**53**:3139-3149. DOI: 10.1016/j.corsci.2011.05.057

[32] Morishige T, Doi H, Goto T, Nakamura E, Takenaka T. Exfoliation corrosion behavior of cold-rolled Mg-14 mass% Li-1 mass% Al alloy in NaCl solution. Materials Transactions. 2013;**54**:1863-1866. DOI: 10.2320/matertrans.MAW201301

[33] Morishige T, Obata Y, Goto T, Fukagawa T, Nakamura E, Takenaka T. Effect of Al composition on the corrosion resistance of Mg-14 mass% Li system alloy. Materials Transactions. 2016;**57**:1853-1856. DOI: 10.2320/matertrans.M2016247

[34] Atrens A, Liu M, Abidin NIZ. Corrosion mechanism applicable to biodegradable magnesium implants. Materials Science & Engineering B. 2011;**176**(20):1609-1636. DOI: 10.1016/j.mseb.2010.12.017

[35] Shi ZM, Atrens A. An innovative specimen configuration for the study of Mg corrosion. Corrosion Science. 2011;**53**:226-246. DOI: 10.1016/j.corsci.2010.09.016

[36] Bobby Kannan M, Dietzel W. Pitting-induced hydrogen embrittlement of magnesium–aluminium alloy. Materials & Design. 2012;**42**:321-326. DOI: 10.1016/j.matdes.2012.06.007

[37] Atrens A, Winzer N, Dietzel W. Stress corrosion cracking of magnesium alloys. Advanced Engineering Materials. 2011;**13**:11-18. DOI: 10.1002/adem.200900287

[38] Wang SD, Xu DK, Han EH, Dong C. Stress corrosion cracking susceptibility of a high strength Mg-7%Gd-5%Y-1%Nd-0.5%Zr alloy. Journal of Magnesium & Alloys. 2014;**2**:335-341. DOI: 10.1016/j.jma.2014.11.004

[39] Zeng RC, Dietzel W, Zettler R, Gan WM, Sun XX. Microstructural evolution and delayed hydride cracking of FSW-AZ31 magnesium alloy during SSRT. Transactions of Nonferrous Metals Society of China. 2014;**24**:3060-3069. DOI: 10.1016/S1003-6326(14)63443-9

[40] Calderón JA, Jiménez JP, Zuleta AA. Improvement of the erosion-corrosion resistance of magnesium by electroless Ni-P/Ni(OH)2-ceramic nanoparticle composite coatings. Surface and Coatings Technology. 2016;**304**:167-178. DOI: 10.1016/j.surfcoat.2016.04.063

[41] Partridge PG. Cyclic twinning in fatigued close-packed hexagonal metals. Philosophical Magazine. 1965;**12**:1043-1054. DOI: 10.1080/14786436508228133

[42] Makar GL, Kruger J, Sieradzki K. Stress corrosion cracking of rapidly solidified magnesium-aluminum alloys. Corrosion Science. 1993;**34**:1311-1323. DOI: 10.1016/0010-938X(93)90090-4

[43] Zeng RC. Corrosion and corrosion fatigue in deformed magnesium alloys [doctoral thesis]. Institute of Metal Research, Chinese Academy of Sciences; 2003 (in Chinese)

[44] Zhou LF, Liu ZY, Wu W, Li XG, Du CW, Jiang B. Stress corrosion cracking behavior of ZK60 magnesium alloy under different conditions. International Journal of Hydrogen Energy. 2017;**42**:26162-26174. DOI: 10.1016/j.ijhydene.2017.08.161

[45] Harandi SE, Banerjee PC, Easton CD, Singh Raman RK. Influence of bovine serum albumin in Hanks' solution on the corrosion and stress corrosion cracking of a magnesium alloy. Materials Science & Engineering C Materials for Biological Applications. 2017;**80**:335-345. DOI: 10.1016/j.msec.2017.06.002

[46] Wang BJ, Wang SD, Xu DK, Han EH. Recent progress in fatigue behavior of Mg alloys in air and aqueous media: A review. Journal of Materials Science & Technology. 2017;**33**:1075-1086. DOI: 10.1016/j.jmst.2017.07.017

[47] Zeng RC, Han EH, Liu L, Xu YB, Ke W. Effect of rolled microstructure on fatigue properties of magnesium alloy AM60. Chinese Journal of Materials Research. 2003;**17**:241-246. DOI: 10.3321/j.issn:1005-3093.2003.03.004

[48] Zeng RC, Han EH, Ke W. Fatigue and corrosion fatigue of magnesium alloys. Materials Science Forum. 2005;**488**:721-724. DOI: 10.4028/www.scientific.net/MSF.488-489.721

[49] Zeng RC, Ke W, Han EH. Influence of load frequency and ageing heat treatment on fatigue crack propagation rate of as-extruded AZ61 alloy. International Journal of Fatigue. 2009;**31**:463-467. DOI: 10.1016/j.ijfatigue.2008.07.005

[50] Zeng RC, Xu YB, Ke W, Han EH. Fatigue crack propagation behavior of an as-extruded magnesium alloy AZ80. Materials Science & Engineering A. 2009;**509**:1-7. DOI: 10.1016/j.msea.2009.01.013

[51] Potzies C, Ulrich KK. Fatigue of magnesium alloys. Advanced Engineering Materials. 2004;**6**:281-289. DOI: 10.1002/adem.200400021

[52] Ogarevic VV, Stephens RI. Fatigue of magnesium alloys. Annual Review of Materials Research. 2003;**20**:141-177. DOI: 10.1146/annurev.ms.20.080190.001041

[53] Zeng RC, Ke W, Han EH. Effect of temperature and relative humidity on fatigue crack propagation behavior of AZ61 magnesium alloy. Materials Science Forum. 2007;**546-549**: 409-412. DOI: 10.4028/www.scientific.net/MSF.546-549.409

[54] Sajuri ZB, Miyashita Y, Mutoh Y. Fatigue characteristics of an extruded AZ61 magnesium alloy. Journal of Japan Institute of Light Metals. 2002;**52**:161-166. DOI: 10.2464/jilm.52.161

[55] Zeng RC. Mechanism of corrosion fatigue for as-extruded magnesium alloy AZ80. Chinese Journal of Materials Research. 2004;**18**:561-567. DOI: 1005-3093(2004)06-0561-07 (in Chinese)

[56] Zeng RC, Ke W. Corrosion fatigue of as-extruded AM60 magnesium alloy. Chinese Journal of Materials Research. 2005;**19**:1-7. DOI: 10.3321/j.issn:1005-3093.2005.01.001 (in Chinese)

[57] Eliezer A, Gutman EM, Abramov E, Unigovski Y. Corrosion fatigue of die-cast and extruded magnesium alloys. Journal of Light Metals. 2001;**1**:179-186. DOI: 10.1016/S1471-5317(01)00011-6

[58] Gu XN, Zhou WR, Zheng YF, Cheng Y, Wei SC, Zhong SP, et al. Corrosion fatigue behaviors of two biomedical Mg alloys—AZ91D and WE43—In simulated body fluid. Acta Biomaterialia. 2010;**6**:4605-4613. DOI: 10.1016/j.actbio.2010.07.026

[59] Jafari S, Singh Raman RK, Davies CHJ. Corrosion fatigue of a magnesium alloy in modified simulated body fluid. Engineering Fracture Mechanics. 2015;**137**:2-11. DOI: 10.1016/j.engfracmech.2014.07.007

[60] Bian D, Zhou WR, Liu Y, Li N, Zheng YF, Sun ZL. Fatigue behaviors of HP-Mg, Mg–Ca and Mg–Zn–Ca biodegradable metals in air and simulated body fluid. Acta Biomaterialia. 2016;**41**:351-360. DOI: 10.1016/j.actbio.2016.05.031

[61] Harandi SE, Raman RKS. Corrosion fatigue of a magnesium alloy under appropriate human physiological conditions for bio-implant applications. Engineering Fracture Mechanics. 2017;**186**:134-142. DOI: 10.1016/j.engfracmech.2017.09.031

[62] Khan SA, Bhuiyan MS, Miyashita Y, Mutoh Y, Koike T. Corrosion fatigue behavior of die-cast and shot-blasted AM60 magnesium alloy. Materials Science & Engineering A. 2011;**528**:1961-1966. DOI: 10.1016/j.msea.2010.11.033

[63] Jafari S, Singh Raman RK. In-vitro biodegradation and corrosion-assisted cracking of a coated magnesium alloy in modified-simulated body fluid. Materials Science & Engineering C. 2017;**78**:278. DOI: 10.1016/j.msec.2017.04.079

[64] Huang WJ, Hou B, Pang YX, Zhou ZR. Fretting wear behavior of AZ91D and AM60B magnesium alloys. Wear. 2006;**260**:1173-1178. DOI: 10.1016/j.wear.2005.07.023

[65] Zhang XP, Zhao ZP, Wu FM, Wang YL, Wu J. Corrosion and wear resistance of AZ91D magnesium alloy with and without microarc oxidation coating in Hank's solution. Journal of Materials Science. 2007;**42**:8523-8528. DOI: 10.1007/s10853-007-1738-z

[66] Chen J, Zeng R-C, Huang W-J, Zheng Z-Q, Wang Z-L, Wang J. Characterization and wear resistance of macro-arc oxidation coating on magnesium alloy AZ91 in simulated body fluids. Transactions of Nonferrous Metals Society of China. 2008;**18**:s361-s364. DOI: 10.1016/S1003-6326(10)60232-4

[67] Wang J, Giridharan V, Shanov V, Xu Z, Collins B, White L, et al. Flow-induced corrosion behavior of absorbable magnesium-based stents. Acta Biomaterialia. 2014;**10**:5213-5223. DOI: 10.1016/j.actbio.2014.08.034

[68] Yu XW, Jiang B, Yang H, Yang QS, Xia XS, Pan FS. High temperature oxidation behavior of Mg-Y-Sn, Mg-Y, Mg-Sn alloys and its effect on corrosion property. Applied Surface Science. 2015;**353**:1013-1022. DOI: 10.1016/j.apsusc.2015.07.011

[69] Song YW, Shan DY, Chen RS, Han E-H. Effect of second phases on the corrosion behaviour of wrought Mg–Zn–Y–Zr alloy. Corrosion Science. 2010;**52**:1830-1837. DOI: 10.1016/j.corsci.2010.02.017

[70] Srinivasan A, Huang Y, Mendis CL, Blawert C, Kainer KU, Hort N. Investigations on microstructures, mechanical and corrosion properties of Mg–Gd–Zn alloys. Materials Science & Engineering A. 2014;**595**:224-234. DOI: 10.1016/j.msea.2013.12.016

Microstructure and Properties of Casting Magnesium Alloys Designed to Work in Elevated Temperature

Andrzej Kiełbus

Additional information is available at the end of the chapter

http://dx.doi.org/10.5772/intechopen.80291

Abstract

Magnesium alloys are widely used in aerospace and automotive industry due to their low density, good mechanical properties, and good castability. Their main disadvantage is low maximum working temperature (about 120°C for Mg-Al alloys). This led to the development of Mg-Al-RE or Mg-RE-Zr alloys, which can work up to 250°C. The chapter will relate to the sand cast and high pressure die cast magnesium alloys. Material for the research consisted of six magnesium casting alloys: AE44, AJ62, WE54, EV31A, and for comparison AZ91 and AM50. The influence of casting and heat treatment parameters on the microstructure and mechanical properties will be introduced. The relationship between the initial structure, casting parameters, phase composition, and mechanical properties in magnesium alloys will be presented.

Keywords: magnesium alloys, sand casting, high-pressure die casting, heat treatment, microstructure, phase composition, mechanical properties

1. Introduction

Due to chemical composition, casting magnesium alloys are divided into two groups [1]. The first group includes alloys containing from 3 to 10% Al with the addition of Zn and Mn. They are characterized by a low cost of manufacture, good tensile strength, elongation, and resistance to atmospheric corrosion. The most popular representative of this group of alloys is AM50 alloy mainly for die casting and AZ91 that can be cast by sand and die casting method. The main advantage of these alloys is their relatively low price, while their disadvantage is low operating temperature—below 120°C [2]. To increase the operating temperature of Mg-Al alloys, alloying elements are introduced, such as rare earth metals, strontium, calcium, and other. Application of rare earth elements to Mg-Al alloys causes a formation of stable thermodynamic phases at the grain boundaries, which provides motivation for improving creep resistance [3]. In particular, the microstructure is connected by the $Al_{11}RE_3$ phase, which is characterized by a high thermodynamic stability to the temperature of ~180°C. The high thermodynamic stability of this phase uses all the aluminum atoms to form this phase during spheroidization, which prevents the $Mg_{17}Al_{12}$ phase from forming. Consequently, these alloys may be applied in the automotive industry for engine elements, gearing, oil pans, and other structural materials working at temperatures of ~180°C [4, 5]. High prices of rare earth elements and their small availability oblige us to search for alternative solutions [6]. Some of those solutions are Mg-Al alloys to which strontium is added. The α-Mg solid solution and eutectic, which consists of one or more intermetallic phases, form the microstructure of Mg-Al-Sr alloys. The Al_4Sr phase is mainly observed in alloys with no more than 5% Al; however, the undesirable $Mg_{17}Al_{12}$ phase can be found in alloys with no more than 6% concentration of Al [7, 8].

The second group includes alloys containing mainly Zn, RE, and Y, without the addition of Al, but always with the addition of Zr. These alloys can be used at a temperature higher than 120°C (up to 250°C), but the price of alloy additions increases the cost of their production. They are mainly used as sand castings. In this group, the most widely used alloys are WE54 and the latest EV31A (Elektron 21) [1]. WE54 magnesium alloy reaches high specific strength, creep resistance, and corrosion resistance up to a temperature of 250°C. The strength of this alloy is achieved essentially via precipitation strengthening. Depending on the aging temperature and time, the precipitating sequence in WE alloys has been reported to involve the formation of phases designated β″, β′, and β [9, 10]. EV31A is magnesium-based sand casting alloy containing neodymium, gadolinium, and zinc for used to approximately 200°C. This alloy has high strength, good corrosion resistance, and excellent castability. In as-cast condition, EV31A alloy is characterized by a solid solution structure α-Mg with eutectic α-Mg + Mg_3(Nd, Gd) intermetallic phase on grain boundaries. Depending on the aging temperature and time, the decomposition of α-Mg supersaturated solid solution is as follows: α-Mg → β″ → β′ → β(Mg_3RE) → $Mg_{41}Nd_5$ [11]. EV31A is being used in both civil and military aircraft and also in the automobile (motorsport) industry [12].

2. Material for research

The material used for the research consisted of the AM50, AZ91, AE44, AJ62, EV31A, and WE54 casting magnesium alloys. The chemical composition of these alloys is provided in **Table 1**.

Alloy	Al	Mn	Zn	Si	Ce	La	Sr	Nd	Gd	Y	Zr	Mg
AM50	4.9	0.45	—		—	—	—	—	—	—	—	Balance
AZ91	8.9	0.24	0.6	0.1	—	—	—	—	—	—	—	
AE44	4.25	0.18	—		2.35	1.07	—	0.59	—	—	—	
AJ62	6.15	0.42	—		—	—	2.1	—	—	—	—	
WE54	—	—	—		—	—	—	1.7	—	5.0	0.55	
EV31A	—	0.001	0.4				—	2.7	1.2	—	0.49	

Table 1. The chemical composition of the investigated magnesium alloys in wt.%.

3. Research methodology

The chemical composition of investigated alloys was measured on the SPEKTROMAX spectrometer. Sand casting was carried out at 700°C (AM50, AZ91, AE44, and AJ62) and 780°C (WE54 and EV31A) temperature. Hot chamber die casting was performed at 650°C (AM50 and AJ62) and 680°C (AE44) temperature. Long-term annealing of AM50 and AZ91 alloys was conducted at two different temperatures: 180 and 250°C for 500–5000 h with air cooling. For AE44 and AJ62 alloys, annealing at temperature 350°C was additionally applied. The as-cast specimens of WE54 and EV31A alloys were solution treated at 520°C for 8 h and quenched into water. Aging treatments were performed at 200, 250, 350, and 420°C for 4–1000 h with air cooling.

For the microstructure observation, an OLYMPUS GX71 metallographic microscope and HITACHI S-3400 N scanning electron microscope were used. TEM examination was carried out on a Tecnai G^2 transmission electron microscope equipped with a high-angle annular dark-field detector (HAADF) and energy-dispersive X-ray (EDX) spectrometer. Metallographic specimens were made in accordance with the methodology developed at the Institute of Materials Engineering at the Silesian University of Technology. X-ray diffraction patterns were collected using an X-Pert Philips diffractometer. Hardness tests have been performed with a Vickers indenter. The examination of the mechanical properties was conducted on an MTS-810 machine at ambient (ca. 20°C) and 200°C.

4. Microstructure and properties of Mg-Al alloys

4.1. AM50 and AZ91 alloys

Sand casting AM50 alloy is characterized by the structure of α-Mg solid solution with precipitates of two types of $Mg_{17}Al_{12}$ phases (**Figure 1a**). First one of massive morphology, together with the solid solution, forms partially divorced eutectic (continuous $Mg_{17}Al_{12}$ + α-Mg) at the grain boundaries. Divorced eutectic $Mg_{17}Al_{12}$ + α-Mg is characterized by the presence of "islands" of α-Mg solid solution, solidifying due to eutectic reaction, which are surrounded by $Mg_{17}Al_{12}$ phase precipitates. The second one of plate morphology is created as a result of discontinuous diffusional transformation. Discontinuous precipitation occurs mostly in the α-Mg regions near the massive $Mg_{17}Al_{12}$ phase on the solid solution α-Mg grain boundaries in regions with higher aluminum. The volume fraction of α-Mg solid solution within $Mg_{17}Al_{12}$ phase precipitates is much smaller than it appears from the balance system. Moreover, globular precipitates of Al_8Mn_5 phase occur in the AM50 alloy.

Casting magnesium alloy AZ91 is the most popular and relatively cheap in comparison with other magnesium alloys available on the market. Aluminum causes an increase in tensile strength and hardness of the alloy but only to a temperature of 120°C. The AZ91 magnesium alloy, like AM50 alloy in as-cast condition, was characterized by a solid solution α-Mg with discontinuous and continuous precipitates of the $Mg_{17}Al_{12}$ phase (**Figure 1b**), with the difference that the volume fraction of the $Mg_{17}Al_{12}$ phase is higher in the AZ91 alloy. Moreover, the occurrence of Mg_2Si and Al_8Mn_5 phases has been provided.

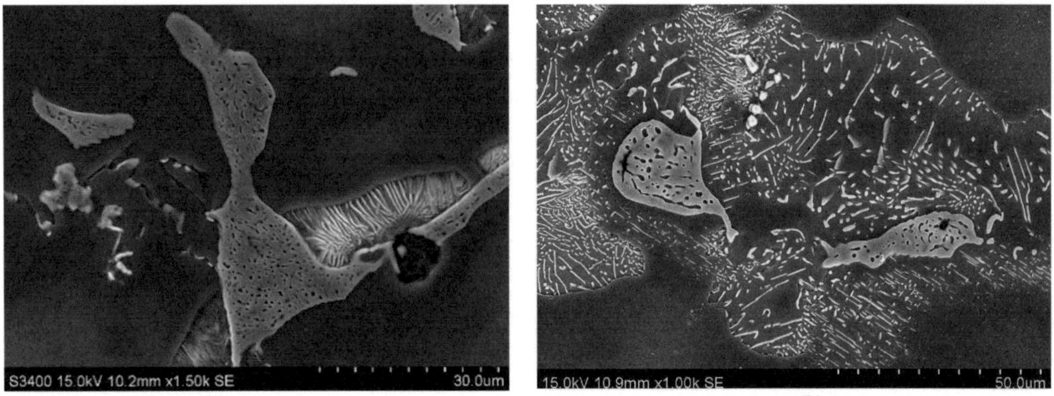

(a) (b)

Figure 1. The microstructure of Mg-Al alloys after sand casting, AM50 (a), AZ91 (b), SEM.

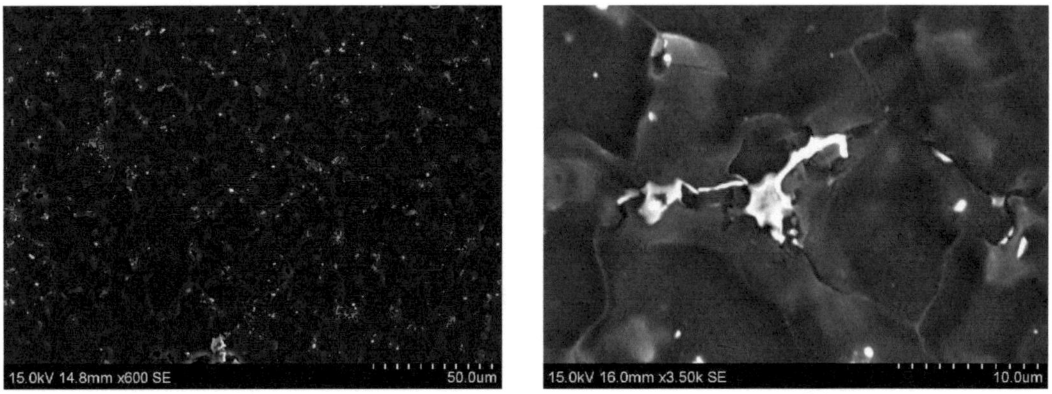

(a) (b)

Figure 2. The microstructure of AM50 alloy after die-casting, SEM.

After die casting, the structure of AM50 alloy is characterized by significant grain refining of the α-Mg solid solution; however, $Mg_{17}Al_{12}$ phase, together with the α-Mg solid solution, forms fully divorced eutectic on the grain boundaries of α-Mg solid solution (**Figure 2**).

During annealing, precipitation of $Mg_{17}Al_{12}$ phase proceeds continuously and discontinuously. In sand casting, firstly, as a result of discontinuous precipitation, precipitates of plate $Mg_{17}Al_{12}$ phase are formed. The process is started on the grain boundaries of the α-Mg solid solution and consists of cellular growth of plate precipitation of $Mg_{17}Al_{12}$ phase in the direction of the central part of the solid solution grain. The growth of lamellar precipitates runs continuously until the alloy matrix reaches the equilibrium composition. The volume fraction of lamellar areas and the distance between the lamellas increases with the rise of temperature, aging time, and an aluminum content in the supersaturated areas. The second step is the coagulation of plate precipitates and the beginnings of continuous precipitation of $Mg_{17}Al_{12}$ phase in zones of increasing content of aluminum. Further annealing causes growth and coagulation of both types of precipitates (**Figure 3**). The extension of annealing time to 5000 h or increasing temperature of annealing to the 250°C causes continuous growing and coagulation of precipitation of $Mg_{17}Al_{12}$ phase (**Figure 4**).

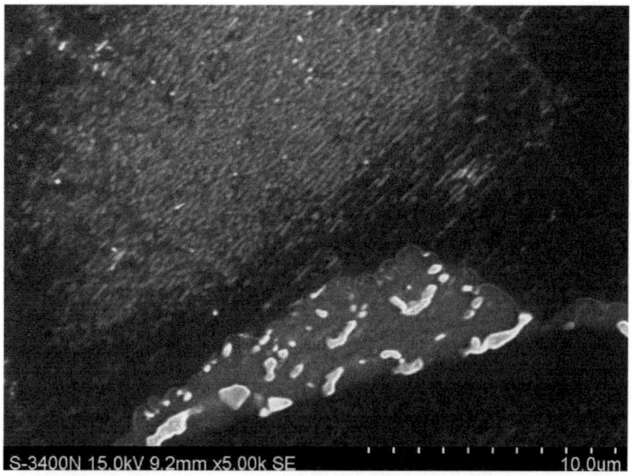

Figure 3. The zones of continuous and coalesced precipitates of $Mg_{17}Al_{12}$ phase in AM50 alloy, after annealing at 180°C/500 h/air.

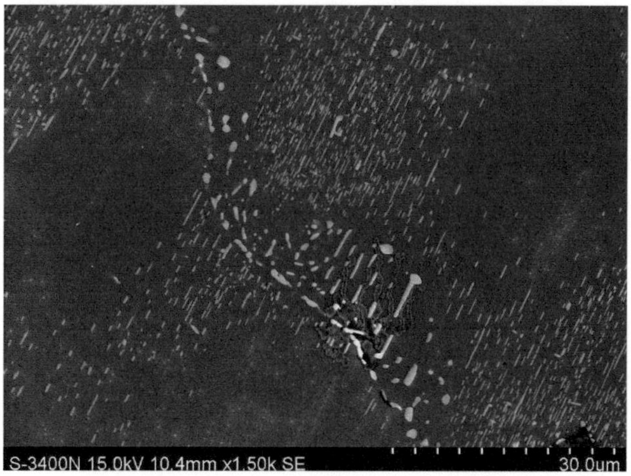

Figure 4. The microstructure of sand casting AM50 alloy, after annealing at 250°C/5000 h/air.

The precipitation processes during long-term annealing in die casting AM50 magnesium alloy proceed similarly as in sand casting, but the difference is that discontinuous precipitation of $Mg_{17}Al_{12}$ phase is not observed in the first step of the process. After 4000 h of annealing, a big, coagulated precipitation of $Mg_{17}Al_{12}$ phase makes the structure of the alloy on the grain boundaries of α-Mg solid solution (**Figure 5**). Long-term annealing at 250°C temperature causes growth and coagulation of continuous precipitates of $Mg_{17}Al_{12}$ phase (**Figure 6**).

The precipitation processes of the $Mg_{17}Al_{12}$ phase during long-term annealing of sand casts resulted in the growth of the hardness of the alloy alongside with lengthening the annealing time. However, in die casts, the structure of the alloy undergoes degradation in the result of precipitation and the coagulation of $Mg_{17}Al_{12}$ phase, what reduces hardness significantly (**Figure 7**).

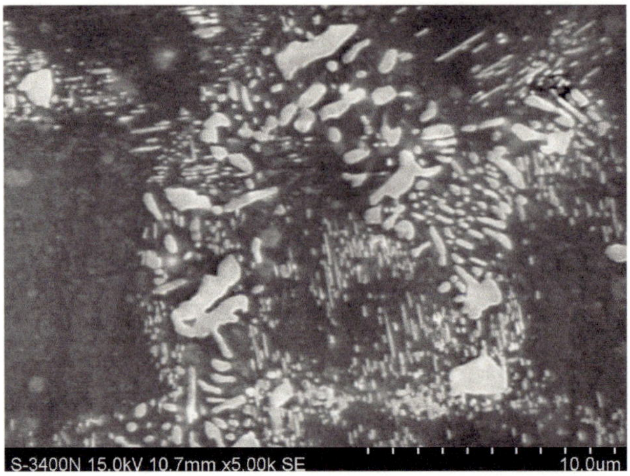

Figure 5. Coalesced continuous precipitates of $Mg_{17}Al_{12}$ phase in die-casting AM50 alloy after annealing at 180°C/4000 h/air.

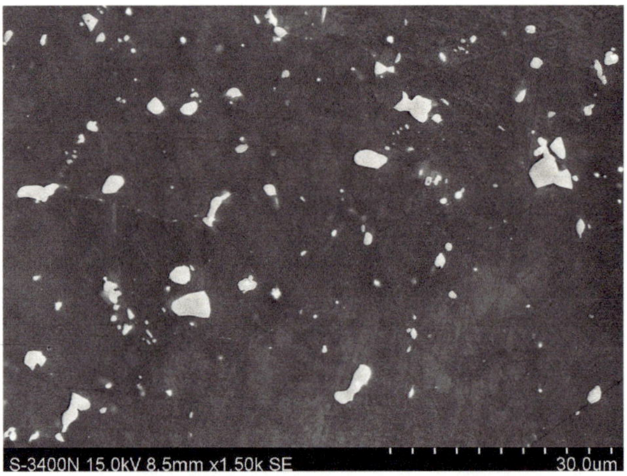

Figure 6. Coagulation of continuous precipitates of $Mg_{17}Al_{12}$ phase in die-casting AM50 alloy after annealing at 250°C/4000 h/air.

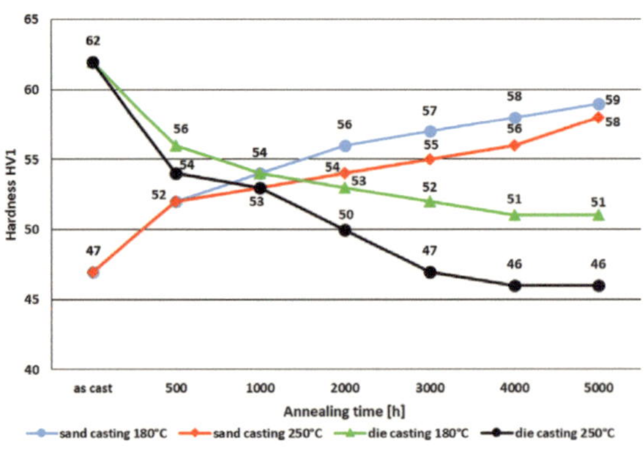

Figure 7. Influence of temperature and annealing time on the hardness of the AM50 alloy.

AM50 and AZ91 alloys due to very good mechanical properties at an ambient temperature and low at elevated temperature can only be used at up to 120°C. This is connected with the presence of $Mg_{17}Al_{12}$ phase, which helps to increase the tensile strength at an ambient temperature but decreases the mechanical properties at an elevated temperature.

5. Microstructure and properties of Mg-Al-X (RE, Sr) alloys

5.1. AE44 alloy

Addition of rare earth elements (mainly cerium, lanthanum, and neodymium) to an alloy with 4% aluminum increases durability and operating temperature of the alloy. The microstructure of the AE44 alloy after sand casting is dominated by the $Al_{11}RE_3$ phase. Moreover, the Al_2RE and $Al_{10}RE_2Mn_7$ phases occur (**Figure 8a**). Die-casting causes grain refining of the solid solution. Diverse morphology of the interdendritic phases occurs at the α-Mg grain boundaries. Needle precipitates of the $Al_{11}RE_3$ phase together with the α-Mg solid solution form a eutectic like–fiber morphology. In contrast to the situation for sand casting, the metastable $Al_{2.12}RE_{0.88}$ phase forms next to eutectic areas (**Figure 8b**).

The $Al_{11}RE_3$ and Al_2RE compounds are typical in Mg-Al-RE alloys. The $Al_{2.12}RE_{0.88}$ is not an equilibrium phase in the binary Al-RE (Al-La) system. The presence of this compound is a result of rapid crystallization during die casting and that it is a metastable phase at room temperature. The $Al_{2.12}RE_{0.88}$ compound was not found in slowly cooled sand cast AE44 alloy.

After long-term annealing at 180°C for 3000 h, the structure of AE44 alloy reveals no significant changes. Continuous precipitation of the $Mg_{17}Al_{12}$ phase is observed only in aluminum supersaturated areas of the α-Mg solid solution in the sand cast alloy (**Figure 9**). The morphology of $Mg_{17}Al_{12}$ phase suggests continuous precipitation from a supersaturated solid solution. These precipitates were occasionally observed in the alloy. Increasing the annealing temperature to 250°C accelerates the process of fragmentation and spheroidization of $Al_{11}RE_3$ phase precipitates (**Figure 10**).

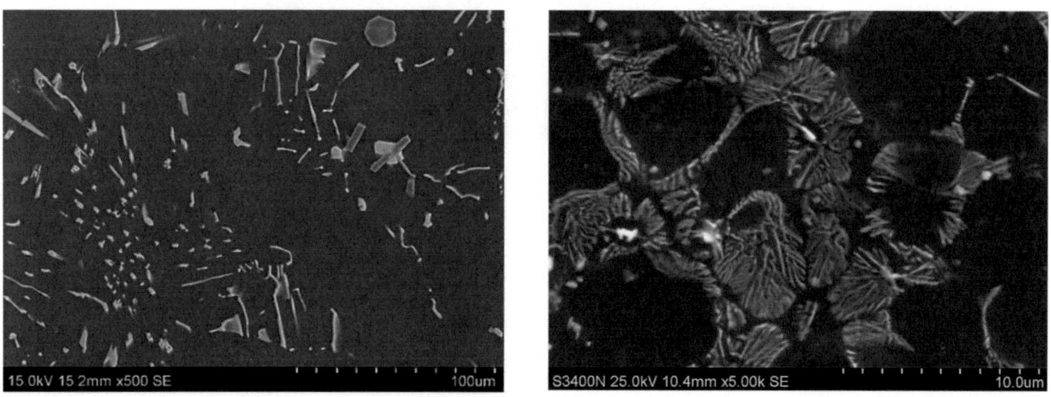

(a) (b)

Figure 8. The microstructure of the AE44 alloy after sand casting (a), die casting (b), SEM.

Figure 9. Precipitates of $Mg_{17}Al_{12}$ phase in the sand cast AE44 alloy after annealing at 180°C/3000 h.

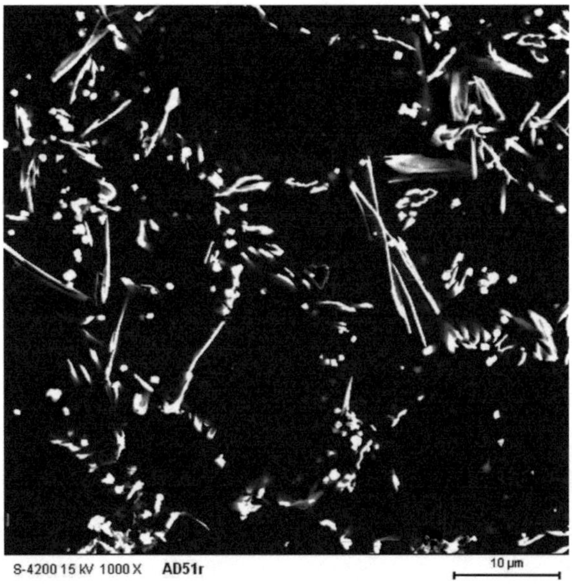

Figure 10. Fragmentation of $Al_{11}RE_3$ phase precipitates in the die-cast AE44 alloy after annealing at 250°C/3000 h.

During long-term annealing of die-cast AE44 alloy, the metastable $Al_{2.12}RE_{0.88}$ phase undergoes a transition into the equilibrium Al_2RE phase (**Figure 11**).

An increase in the hardness (only sand cast) after annealing at the temperature of 180°C is caused by the separation of $Mg_{17}Al_{12}$ phase in the areas of increased aluminum content (**Figure 12a**). The fragmentation and spheroidization of $Al_{11}RE_3$ phase after annealing at

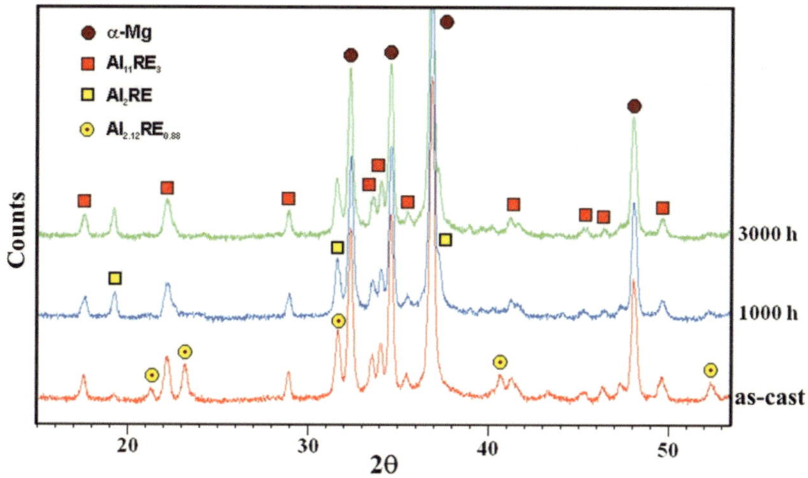

Figure 11. X-ray diffractions of the die-casting AE44 alloy after annealing for 1000 and 3000 h at the temperature of 180°C.

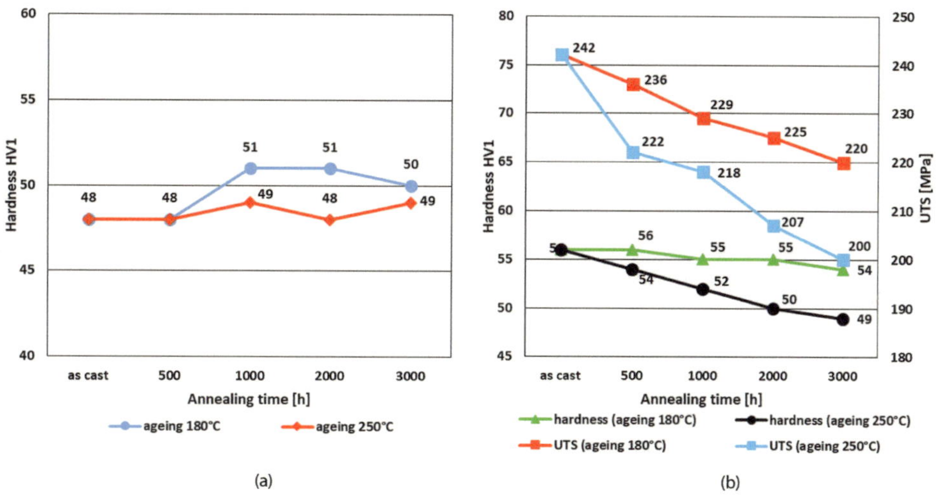

Figure 12. The influence of temperature and annealing time on the hardness and tensile strength of the AE44 alloy after sand casting (a) and die casting (b).

temperatures of 180 and 250°C caused a significant decrease in hardness and tensile strength of die casting AE44 (**Figure 12b**).

5.2. AJ62 alloy

The AJ62 alloy is characterized by the structure of the α-Mg solid solution with precipitates of intermetallic phases of type: (Al, Mg)$_4$Sr, $Al_3Mg_{13}Sr$, and Mn_5Al_8. The (Al, Mg)$_4$Sr phase and the solid solution form eutectic of α-Mg + (Al, Mg)$_4$Sr. However, the $Al_3Mg_{13}Sr$ phase occurs at the grain boundaries of the α-Mg solid solution in the direct surroundings of eutectic areas. Globular precipitates of the Mn_5Al_8 phase occur at grain boundaries and inside the grains of

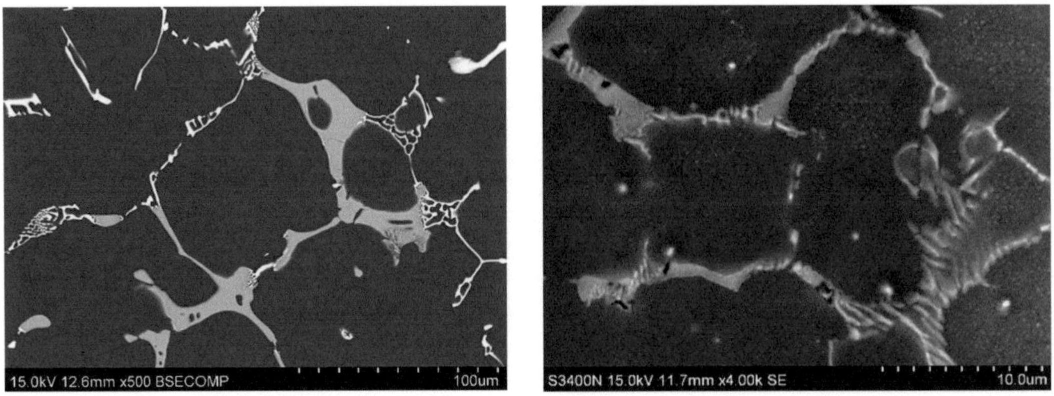

(a) (b)

Figure 13. The microstructure of AJ62 alloy after sand casting (a), die casting (b), SEM.

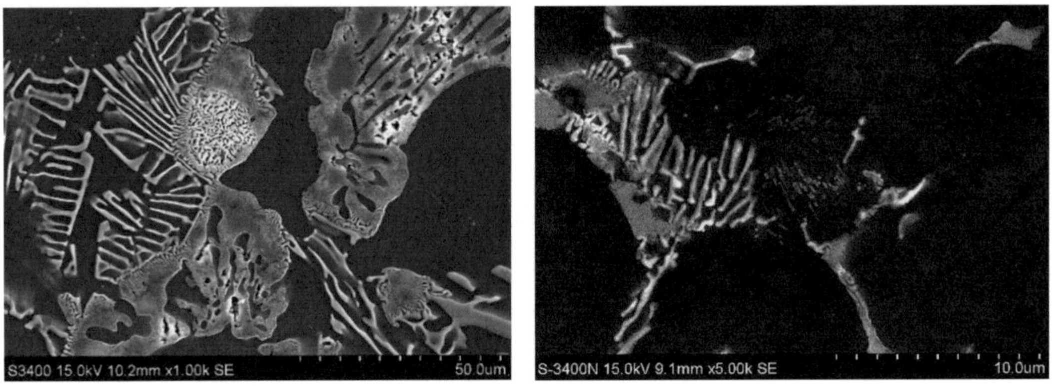

(a) (b)

Figure 14. Precipitates of $Mg_{17}Al_{12}$ and initial decomposition of $Al_3Mg_{13}Sr$ phase after annealing of sand casting for 500 h at 250°C (a) and die casting for 4000 h at 180°C (b).

the α-Mg solid solution. In the thin-wall die casts, eutectic α-Mg + $(Al, Mg)_4Sr$ occurs at the grain boundaries. Moreover, for thick-wall casting, which is characterized by a lower cooling rate, the $Al_3Mg_{13}Sr$ phase occurs (**Figure 13**).

The applied casting technology has no influence on the type of changes in the AJ62 alloy microstructure during long-term annealing. The first step occurs at ~180°C (250°C/500 h) and is characterized by precipitation of the $Mg_{17}Al_{12}$ phase in areas of higher aluminum volume (as in AE44 alloy) and the beginning of $Al_3Mg_{13}Sr$ phase decomposition (**Figure 14**).

Decomposition of this phase results in the formation of plate precipitates of the $(Al, Mg)_4Sr$ phase separated by areas of α-Mg solid solution. The second step occurs at ~250°C and consists of complete degradation of the $Al_3Mg_{13}Sr$ phase by the reaction $Al_3Mg_{13}Sr \rightarrow (Al, Mg)_4Sr + \alpha\text{-}Mg$ (**Figure 15**). The third step occurs above 300°C. In the α-Mg solid solution,

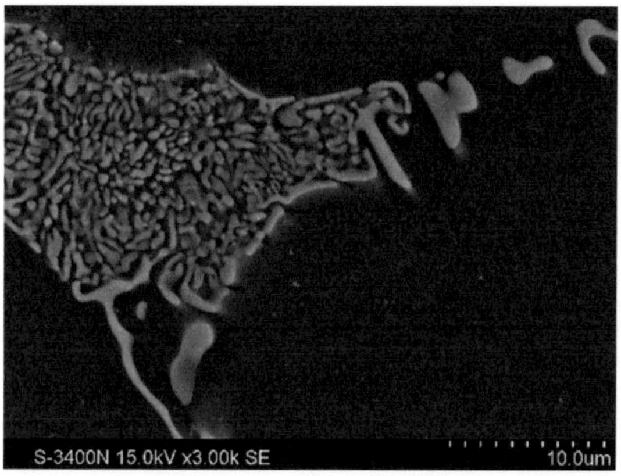

Figure 15. Complete decomposition of $Al_3Mg_{13}Sr$ phase in sand-cast AJ62 alloy after annealing for 1000 h at 350°C, SEM.

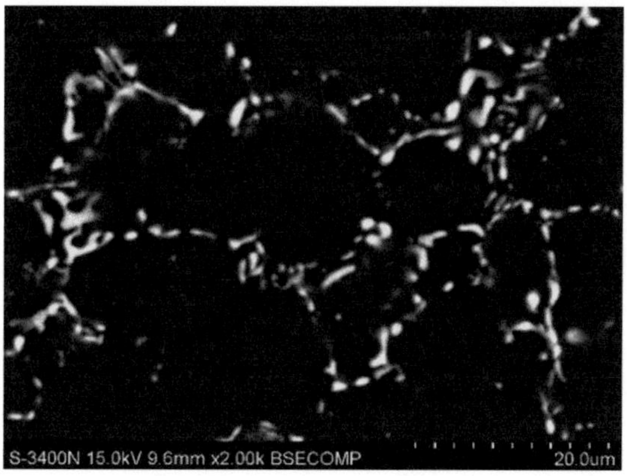

Figure 16. Coagulation of (Al, Mg)$_4$Sr phase precipitates in die-cast AJ62 alloy after annealing for 4000 h at 350°C, SEM.

there are only precipitates of the primary and secondary (generated from $Al_3Mg_{13}Sr$ phase decomposition) (Al, Mg)$_4$Sr phase. Long-term annealing at this temperature leads to fragmentation and coagulation of (Al, Mg)$_4$Sr phase precipitates (**Figure 16**).

The hardness and tensile strength evolution of AJ62 alloy as a function of aging time for isothermal aging at 200, 250, and 350°C is shown in **Figure 17**.

Increase in hardness after annealing at the temperature of 180°C as well as in the first stage of annealing at the temperature of 250°C (only sand cast) is caused by the separation of $Mg_{17}Al_{12}$ phase in the areas of increased aluminum content. The hardness of the die cast, unlike of sand cast, does not decrease after annealing at the temperature of 250°C. It is related to the lower number of $Al_3Mg_{13}Sr$ phase precipitates. The decomposition of $Al_3Mg_{13}Sr$ phase and coagulation of (Al, Mg)$_4$Sr phase precipitates after annealing at the temperature of 350°C cause a significant decrease in hardness and tensile strength of AJ62 alloy, regardless of applied casting technology.

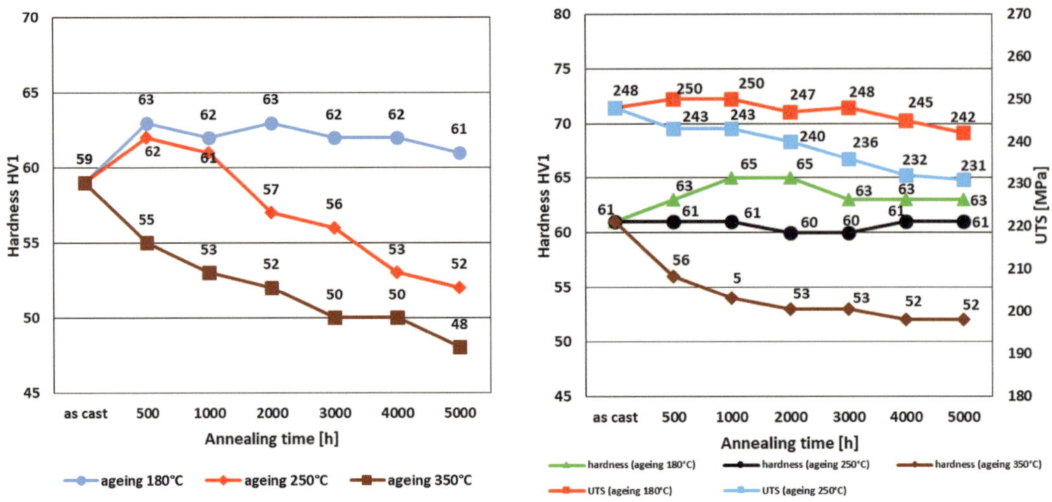

(a) (b)

Figure 17. The influence of temperature and annealing time on the hardness and tensile strength of the AJ62 alloy after sand casting (a) and die casting (b).

6. Microstructure and properties of Mg-Zr-X alloys

6.1. WE54 alloy

The WE54 alloy in as-cast condition was characterized by a solid solution structure α-Mg with eutectic α-Mg + β on grain boundaries (**Figure 18**). Equilibrium β phase is isomorphic to the Mg_5Gd phase and is identified as a $Mg_{14}Nd_2Y$ phase. Moreover, the occurrence of MgY, Mg_2Y, and $Mg_{24}Y_5$ phases has been provided [13]. Also, the zirconium-rich core areas have been observed. Zirconium-rich core areas are ellipsoidal or nearly circular (**Figure 19**).

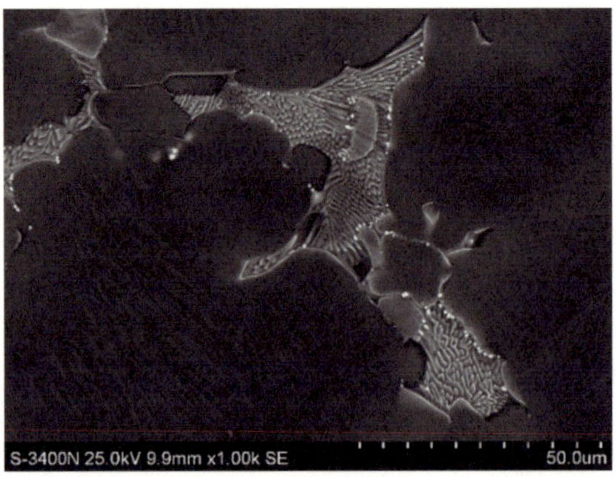

Figure 18. The WE54 alloy microstructure in as-cast condition, SEM.

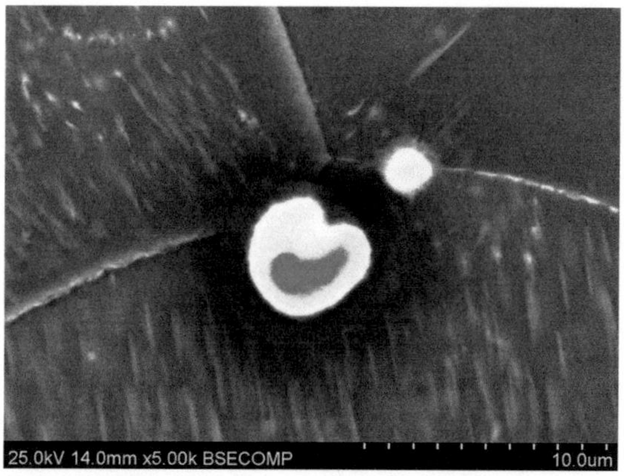

Figure 19. Zirconium-rich core areas in the WE54 alloy.

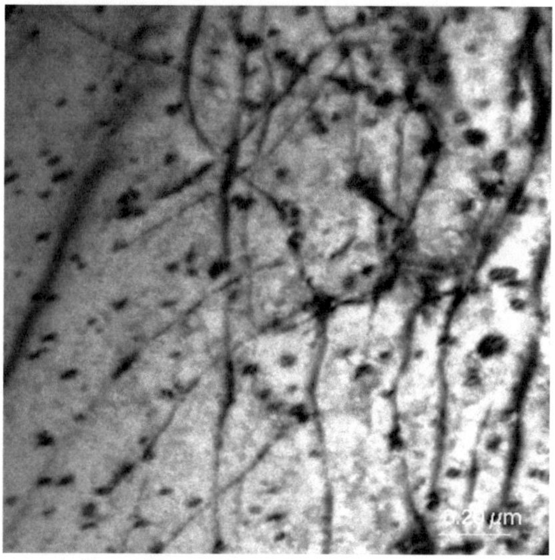

Figure 20. Fine-dispersed precipitates of β″ and β′ phases in the WE54 alloy after aging at 250°C/4 h, TEM.

The high tensile strength of WE54 alloy is related to dispersive precipitations in the microstructure. After 4 hours of aging at 250°C two types of metastable phases, coherent or semi-coherent with the matrix were observed. The first type of precipitates is, coherent with the matrix, β″ phase (**Figure 20**), characterized by type DO_{19} structure [14]. The second phase is, semi-coherent with the matrix β′ phase, characterized by an orthorhombic, space-centered structure [15]. Extension of the aging time up to 16 hours (T6 treatment) causes the disappearance of β″ phase. The β′ phase grows and changes its shape from spherical to lamellar one (**Figure 21**). Extension of the aging time up to 48 h causes the growth and the change of shape of the β′ phase precipitates from spheroidal into lamellar (**Figure 22**) and the beginnings of the phase transformation $\beta' \rightarrow \beta_1$. The resulting new phase, identified as β_1 phase, nucleates heterogeneously on the β′ phase precipitates. The β_1 phase is characterized by a

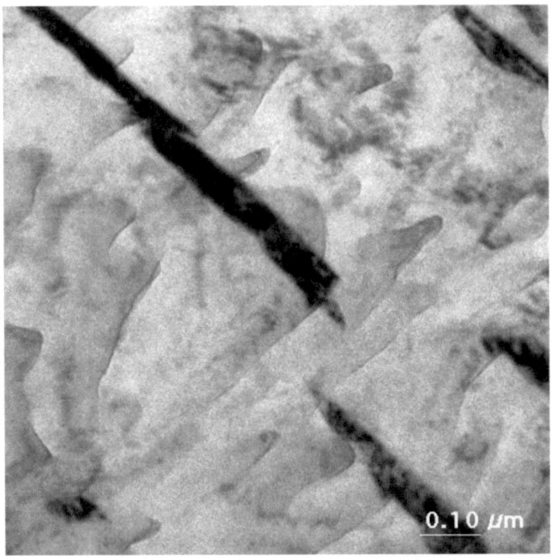

Figure 21. β′ phase precipitates in the WE54 alloy after aging at 250°C/16 h, TEM.

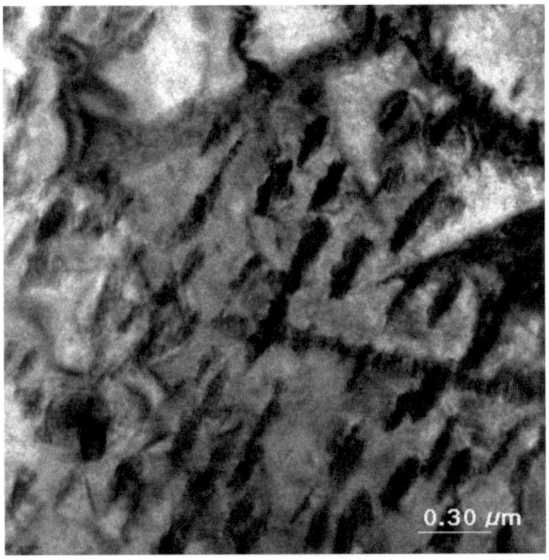

Figure 22. β′ phase precipitates in the WE54 alloy after aging at 250°C/48 h, TEM.

face-centered cubic structure [9]. Further aging (96 h) increases the amount of β_1 phase and causes the emergence of equilibrium β phase (**Figure 23**). Precipitation processes end with the emergence of the equilibrium β phase. It is formed during aging for a long time at 250°C and above (**Figure 24**).

The evolution of hardness and tensile strength as a function of aging time for isothermal aging at 200 and 250°C is shown in **Figure 25**. In practice, for this alloy, aging at 250°C for 16 h is applied (T6 treatment), allowing to obtain maximum tensile strength, which is related to the presence of β″ and β′ phases precipitates in the structure.

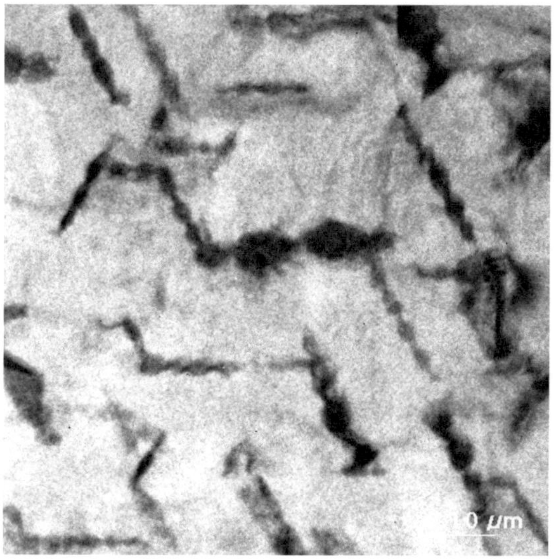

Figure 23. β′ and β1 phases in the WE54 alloy after aging at 250°C/96 h, TEM.

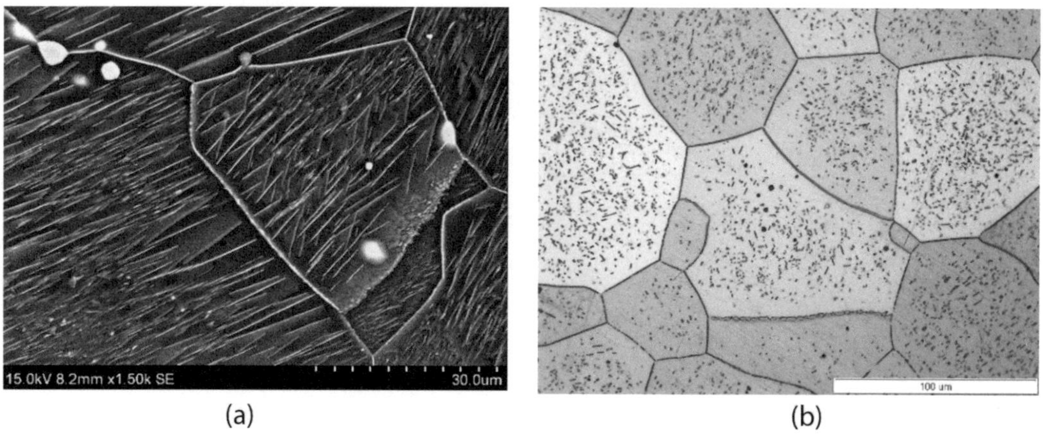

Figure 24. Equilibrium β phase in the WE54 alloy after aging at 300°C/96 h—SEM (a) and 420°C/22 h—LM (b).

The Mg-Y-Zr (WE43 and WE54) alloys are widely used in aircraft and automotive industries. Due to the presence of yttrium, the Mg-Y-Zr alloy belongs to the most expensive of magnesium alloys. This is the main reason for searching other alloys that will fulfill the performance requirements at lower manufacturing costs. One of them is EV31A (Elektron 21) alloy.

6.2. EV31A (Elektron 21) alloy

The EV31A alloy is characterized by the α-Mg solid solution structure with eutectic α-Mg + Mg_3(Nd, Gd) on the grain boundaries and regularly shaped precipitates of a Mg_3Gd phase (**Figure 26**). The Mg_3(Nd, Gd) phase is a modification of Mg_3Nd phase with neodymium substituted by gadolinium without destroying the crystal structure, due to the reasonably small difference in the atomic radii between gadolinium r_{Gd} = 0.1802 nm and neodymium r_{Nd} = 0.1821 nm. Phases occurring in the EV31A alloy take the forms (Mg, Zn)$_3$(Nd, Gd) and Mg(Gd, Nd)$_3$.

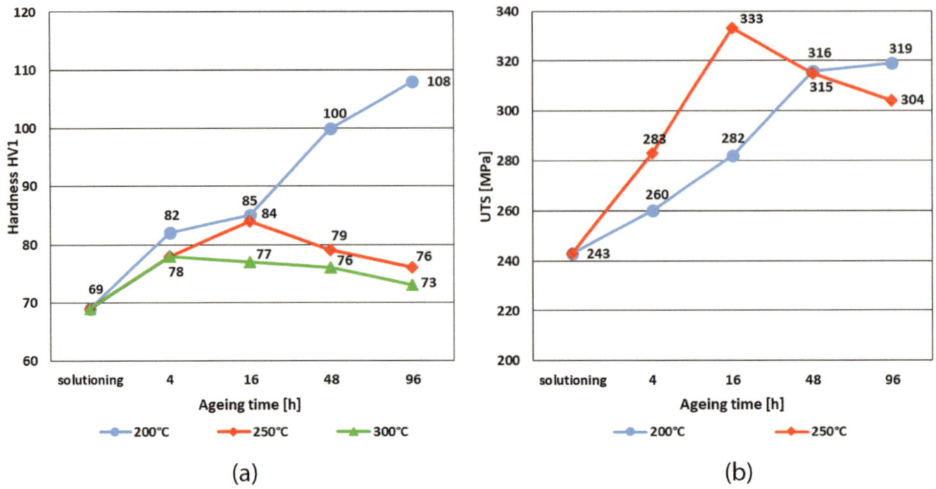

Figure 25. The influence of temperature and annealing time on the hardness (a) and tensile strength (b) of the WE54 alloy.

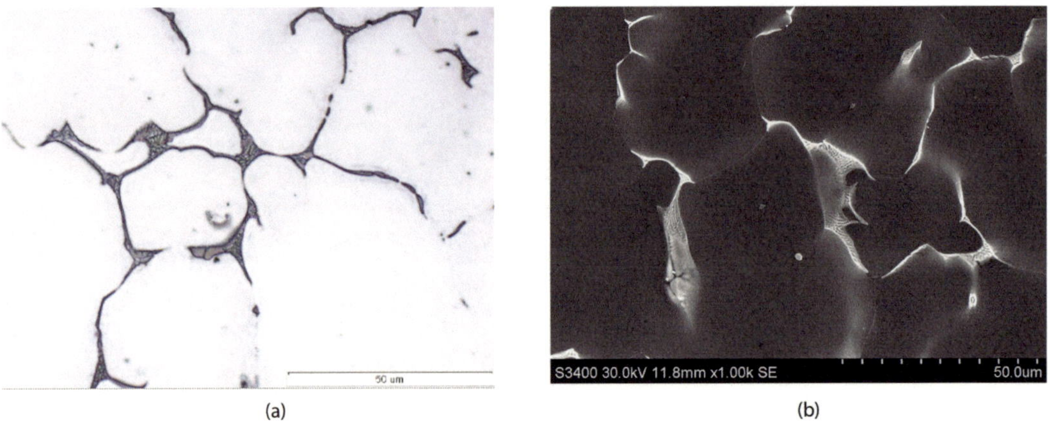

Figure 26. The EV31A alloy microstructure in as-cast condition, LM (a) and SEM (b).

Similarly, as in the WE54 alloy, high tensile strength is connected with dispersive precipitates produced during aging. In the first stage of the aging process (200°C/4 h and 16 h), the product of the α-Mg decomposition is fully coherent and semi-coherent with the matrix precipitates of β″ and β′ phases, respectively (**Figure 27**).

The β″ phase is characterized by the lattice type DO19 (a = 0.64 nm, c = 0.52 nm), while β′ has a face-centered cubic structure (a = 0.72 nm) [16]. The second stage (200°C/48H) is the formation of a stable equilibrium β(Mg_3Nd) phase. The β phase is non-coherent with the α-Mg matrix. The β phase precipitation process is similar to β_1 phase precipitation in the WE54 alloy. In the immediate vicinity of the β′ phase (**Figure 28**), lamellar precipitates of the equilibrium β phase

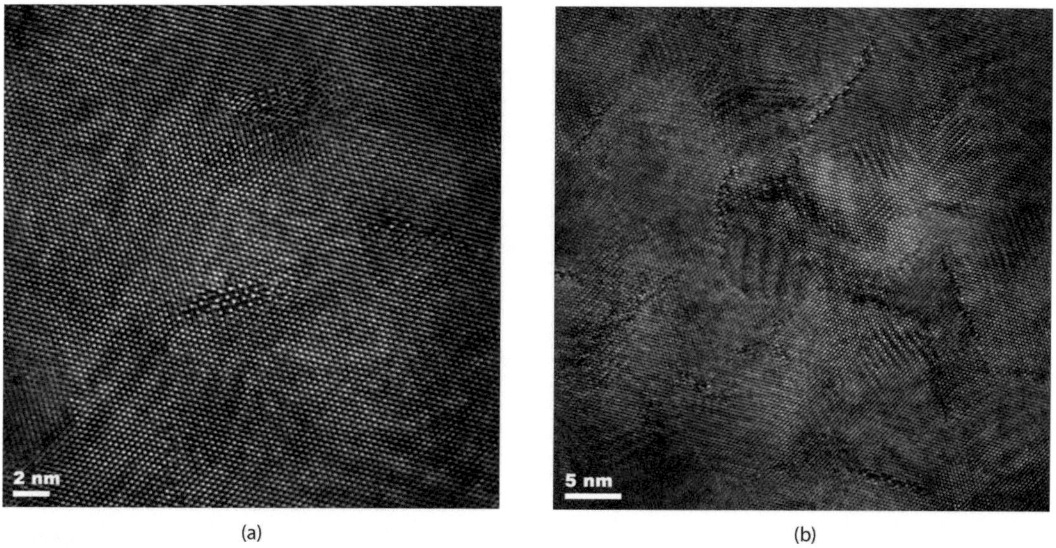

Figure 27. β″ phase precipitates coherent with matrix and separate β′ phase precipitates in EV31A alloy after aging at 200°C/4 h (a) and 200°C/14 h (b), HRTEM image.

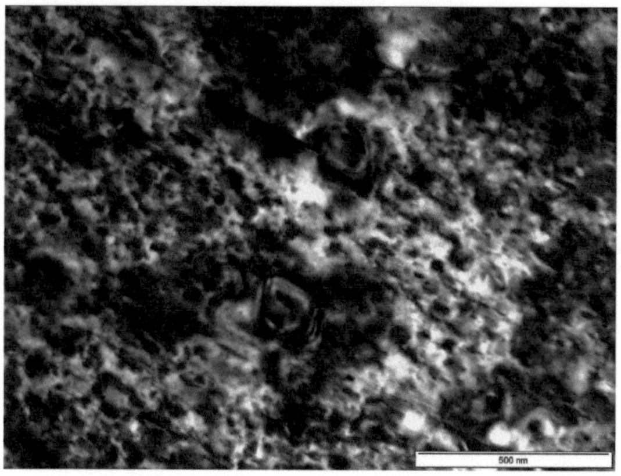

Figure 28. β′ and β phases in the EV31A alloy after aging at 200°C/48 h, TEM.

arise. This indicates that the resulting new phase nucleates heterogeneously at the β′ phase precipitates. After 96 h of aging, the volume fraction of equilibrium β phase increases. The β phase is incoherent with the matrix and identified as an Mg_3Nd phase with face-centered cubic structure (a = 0.7410 nm). On the other hand, aging of this alloy at 300°C leads to the appearance in the structure, next to the lamellar β phase precipitates, of $Mg_{41}Nd_5$ phase precipitates at the solid solution grain boundaries (**Figure 29**).

In the EV31A alloy, after long-term aging at a temperature 350°C/500–5000 h/air, there are no precipitates of equilibrium β phase (**Figure 30**).

Figure 29. β and Mg41Nd5 phases in the EV31A alloy after aging at 250°C/48 h, TEM.

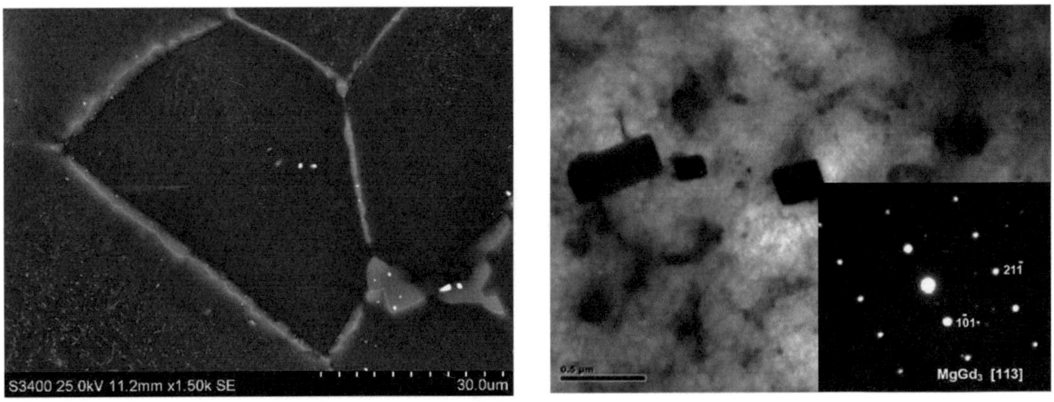

(a) (b)

Figure 30. The microstructure of EV31A alloy after aging at 350°C/1000 h: (a) $Mg_{41}Nd_5$ phase precipitates network at α-Mg solid solution grain boundaries, SEM; (b) TEM image and electron diffraction pattern of MgGd3 phase.

$Mg_{41}Nd_5$ phase creates a characteristic network at the grain boundaries of the α-Mg solid solution. In its vicinity, there are regular $Mg(Nd, Gd)_3$ phase precipitates.

Generally, the investigated alloy showed that the decomposition of α-Mg supersaturated solid solution with increasing aging time is as follows (Eq.(1)):

$$\alpha-\mathrm{Mg} \rightarrow \beta'' \rightarrow \beta' \rightarrow \beta(\mathrm{Mg_3RE}) \rightarrow \mathrm{Mg_{41}Nd_5}. \quad (1)$$

The evolution of hardness and tensile strength as a function of aging time for isothermal aging at 200, 250, and 300°C is shown in **Figure 31**. Alloy showed a remarkable hardening at 200°C temperature, and the peak hardness and good tensile strength can achieve after aging 200°C/16 h due to the precipitation of β″ and β′ phases. The peak hardness was shortened with an increase of the aging temperature. There was not any peak hardness in case of an alloy aged at 300°C. The hardness of EV31A alloy decreases to 45 HV after 500 h of annealing due to precipitation of β and $Mg_{41}Nd_5$ phases.

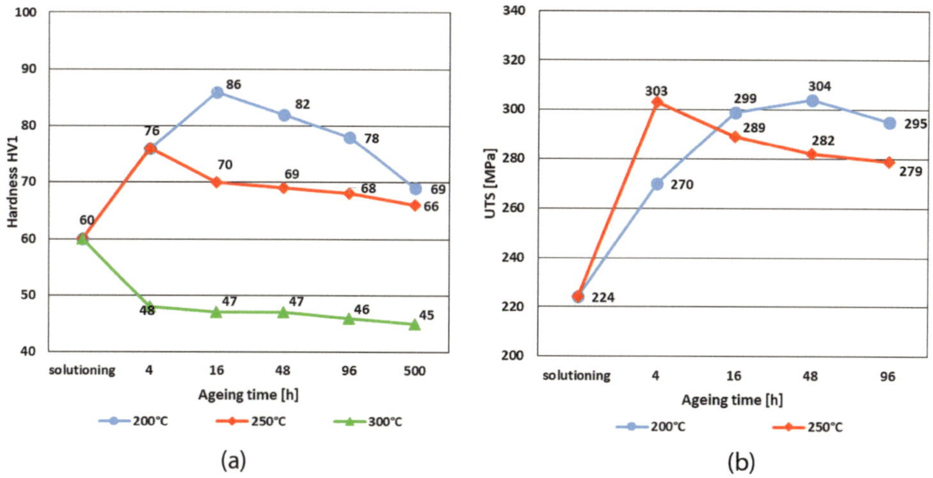

Figure 31. The influence of temperature and annealing time on the hardness (a) and tensile strength (b) of the EV31A alloy.

In practice, for this alloy, aging at 200°C for 16 h is applied (T6 treatment), allowing to obtain maximum tensile strength, which is related to the presence of β″ and β′ phases precipitates in the structure.

7. Conclusions

Results of the research can help in creating conclusions of experience feature, as follows:

1. The AM50 and AZ91 alloys are characterized by the structure of α-Mg solid solution with discontinuous and continuous precipitates of $Mg_{17}Al_{12}$ phase at grain boundaries. The mechanism of precipitation, as well as the volume fraction, morphology, and distribution of $Mg_{17}Al_{12}$ phase, is dependent on temperature and aging time. The $Mg_{17}Al_{12}$ phase undergoes decomposition and coagulation at the temperature above 180°C and higher. The precipitation and degradation processes of this phase in sand-cast result in an increase the hardness of the alloys alongside with lengthening the annealing time. However, in die casts, the coagulation of $Mg_{17}Al_{12}$ phase reduces the hardness of the alloys. The presence of $Mg_{17}Al_{12}$ phase in the structure limits the use of this alloy to a temperature no higher than 120°C.

2. Addition of REs to Mg-Al alloy causes the following phases to form: $Al_{11}RE_3$, Al_2RE, $Al_{2.12}RE_{0.88}$, and $Al_{10}RE_2Mn_7$ in the matrix of the α-Mg solid solution. The microstructure of AE44 alloy during long-term annealing proceeds via two steps. The first step occurs at ~180°C and is characterized by precipitation of the $Mg_{17}Al_{12}$ phase (mainly for sand casting) in areas of higher aluminum volume and by the appearance of changing $Al_{2.12}RE_{0.88} \rightarrow Al_2RE$ (for die casting). The process of precipitation of the $Mg_{17}Al_{12}$ phase is only the result of continuous precipitation in areas of higher aluminum volume. The $Al_{11}RE_3$ phase is stable up to 250°C. The second step stage takes place at a temperature equal to or higher than 250°C and is characterized by significant degradation of the structure, which is manifested by fragmentation and spheroidization of the $Al_{11}RE_3$ phase. The degradation of $Al_{11}RE_3$ phase caused a significant decrease in hardness and tensile strength of die casting AE44.

3. The AJ62 alloy is characterized by the structure of the α-Mg solid solution with precipitates of intermetallic phases of type: $(Al, Mg)_4Sr$, $Al_3Mg_{13}Sr$, and Mn_5Al_8. The $(Al, Mg)_4Sr$ phase and the solid solution form eutectic of α-Mg + $(Al, Mg)_4Sr$. Long-term annealing of this alloy at temperatures 180 and 250°C causes decomposition of $Al_3Mg_{13}Sr$ phase according to the reaction $Al_3Mg_{13}Sr \rightarrow (Al, Mg)_4Sr + \alpha\text{-}Mg$, while at the temperature of 350°C, it leads to fragmentation and coagulation of $(Al, Mg)_4Sr$ phase. Observed microstructure changes cause a decrease of mechanical properties of the alloy.
4. The WE54 alloy is characterized by the structure of the α-Mg solid solution with eutectic $\alpha\text{-}Mg + \beta(Mg_{14}Y_2Nd)$. Moreover, the occurrence of MgY, Mg_2Y, and $Mg_{24}Y_5$ phases has been provided. During aging, the supersaturated magnesium solid solution decomposes in the following sequence $\alpha\text{-}Mg \rightarrow \beta'' \rightarrow \beta' \rightarrow \beta_1 \rightarrow \beta$. The WE54 alloy is widely used in aircraft and automotive industries for components utilized up to a temperature of ~250°C. However, due to its price, there is a search for other alloys, which can fulfill the performance requirements at lower manufacturing costs. One of such alloys is the EV31A alloy.
5. The EV31A alloy is characterized by the α-Mg solid solution structure with the eutectic $\alpha\text{-}Mg + (Mg, Zn)_3(Nd, Gd)$ at the grain boundaries of the α-Mg solid solution and regularly shaped, $MgGd_3$ phase precipitates. The precipitation process during aging runs in accordance with the sequence: $\alpha\text{-}Mg \rightarrow \beta'' \rightarrow \beta' \rightarrow \beta(Mg_3Nd) \rightarrow Mg_{41}Nd_5$. The obtained results indicate that the EV31A alloy can be utilized at temperatures up to ~200°C and can be a substitute for the previously used WE54 alloy.

Acknowledgements

The present work was financed from the research project no. 11/990/BK_18/0057.

Author details

Andrzej Kiełbus

Address all correspondence to: andrzej.kielbus@polsl.pl

Silesian University of Technology, Katowice, Poland

References

[1] Friedrich H, Mordike B. Magnesium Technology. Berlin Heidelberg: Springer-Verlag; 2006. 677 p. DOI: 10.1007/3-540-30812-1

[2] Mordike B, Eberr T. Magnesium: Properties-applications-potentials. Materials Science and Engineering: A. 2001;**302**:37-45. DOI: 10.1016/S0921-5093(00)01351-4

[3] Pettersen G, Westengen H, Høier R, Lohne O. Microstructure of a pressure die-cast magnesium-4wt.% aluminum alloy modified with rare earth additions. Materials Science Engineering: A. 1996;**207**:115-120. DOI: 10.1016/0921-5093(95)10035-0

[4] Bakke P, Westengen H. The role of rare earth elements in structure and property control of magnesium die casting alloys. In: Mathaudhu S, Luo A, Neelameggham N, Nyberg E, Sillekens W, editors. Essential Readings in Magnesium Technology. Cham, Springer International Publishers; 2005. pp. 313-318. DOI: 10.1007/978-3-319-48099-2_50

[5] Powell BR, Rezhets V, Balogh MP, Waldo RA. Microstructure and creep behavior in AE42 magnesium die-casting alloy. The Journal of The Minerals, Metals & Materials Society. 2002;**54**(8):34-38. DOI: 10.1007/BF02711864

[6] Yang M, Pan F, Zhang J. An analysis of the development and applications of current and new Mg-Al based elevated temperature magnesium alloys. Materials Science Forum. 2005;**488-489**:923-926. DOI: 10.4028/www.scientific.net/MSF.488-489.923

[7] Parvez MA, Medraj M, Essadiqi F, Muntasar A, Denes G. Experimental study of the ternary magnesium—Aluminum–strontium system. Journal of Alloys and Compounds. 2005;**402**:170-185. DOI: 10.1016/j.jallcom.2005.04.173

[8] Czerwinski F, Zielinska-Lipiec A. The microstructure evolution during semisolid molding of a creep-resistant Mg-5Al-2Sr alloy. Acta Materialia. 2005;**53**(12):3433-3444. DOI: 10.1016/j.actamat.2005.03.048

[9] Nie JF, Muddle BC. Precipitation in magnesium alloy WE54 during isothermal aging at 250°C. Scripta Materialia. 1999;**40**(10):1089-1094. DOI: 10.1016/S1359-6462(99)00084-6

[10] Rokhlin LL, Dobatkina TV, Tarytina IE, Timofeev VN, Balakhchi EE. Peculiarities of the phase relations in Mg-rich alloys of the Mg-Nd-Y system. Journal of Alloys and Compounds. 2004;**367**:17-19. DOI: 10.1016/j.jallcom.2003.08.004

[11] Kielbus A. Microstructure and properties of elektron 21 magnesium alloy. In: Czerwinski F, editor. Magnesium Alloys. London: IntechOpen; 2011. pp. 281-296. DOI: 10.5772/13220

[12] Lyon P, Syed I, Wilks T. The influence of alloying elements and heat treatment upon properties of Elektron 21 (EV31A) alloy. In: Proceedings of Magnesium Technology—Symposium; 13-17 February 2005; USA. San Francisco: TMS; 2005. pp. 303-308

[13] Kielbus A. Microstructure and properties of sand casting magnesium alloys for elevated temperature applications. Solid State Phenomena. 2011;**176**:63-74. DOI: 10.4028/www.scientific.net/SSP.176.63

[14] Antion C, Donnadieu P, Perrard F, Deschamps A, Tassin C, Pisch A. Hardening precipitation in a Mg–4Y–3RE alloy. Acta Materialia. 2003;**51**:5335-5348. DOI: 10.1016/S1359-6454(03)00391-4

[15] Nie JF, Xiao XL, Luo CP, Muddle BC. Characterisation of precipitate phases in magnesium alloys using electron microdiffraction. Micron. 2001;**32**:857-863. DOI: 10.1016/S0968-4328(00)00094-9

[16] He SM, Zeng XQ, Peng LM, Gao X, Nie JF, Ding WJ. Precipitation in a Mg–10Gd–3Y–0.4Zr (wt.%) alloy during isothermal ageing at 250°C. Journal of Alloys and Compounds. 2006;**421**:309-313. DOI: 10.1016/j.jallcom.2005.11.046

Chapter 5

Superplastic Behaviour of Selected Magnesium Alloys

Zuzanka Trojanová, Zdeněk Drozd and Pavel Lukáč

Additional information is available at the end of the chapter

http://dx.doi.org/10.5772/intechopen.79752

Abstract

Superplastic materials exhibit anomalous plasticity, achieving strain until several thousand per cent. The phenomenon of plasticity is limited on special microstructure, temperatures and strain rates. Magnesium and magnesium alloys are known as materials with limited plasticity. This is due to their hexagonal structure of these materials. Finding the superplasticity conditions has a crucial importance for applications of magnesium alloys. In this chapter, we will deal with the superplastic behaviour of AZ91, QE22, AE42 and EZ33 magnesium alloys. Materials were prepared by various techniques: thermomechanical treatments, equal channel angular pressing, hot extrusion, rolling, friction stirring and high-pressure torsion. Strain rate sensitivity and elongation to fracture were estimated at various temperatures. Mechanisms of superplastic flow are discussed. Grain boundary sliding and diffusional processes were depicted as the main mechanisms responsible for high plasticity of these alloys. On the other hand, cavitation at elevated temperatures deteriorates the superplastic properties.

Keywords: AZ91 magnesium alloy, QE22 magnesium alloy, AE42 magnesium alloy, EZ33 magnesium alloy, superplasticity, strain rate sensitivity, activation energy

1. Introduction

A stable tensile deformation is realised when no localisation (necking) takes place and the cross-sectional area, A, decreases uniformly along the sample length. In such case, the following condition must be fulfilled [1]:

$$\frac{dln\dot{A}}{dlnA} = 0 \tag{1}$$

Materials, which show linear (Newtonian) viscosity, have the strain rate sensitivity parameter m

$$m = \frac{\mathrm{dln}\,\sigma}{\mathrm{dln}\,\dot{\varepsilon}} = 1. \tag{2}$$

Here, σ is the true stress and $\dot{\varepsilon}$ is the strain rate (SR). The materials exhibit no necking and they deform to large extension before fracture. Such behaviour is typical for molten glass and polymers where m approaches unity. Metals and alloys may have similar m values under certain regimes of temperatures and strain rates. This phenomenon of 'unusual' deformation of materials up to several thousand per cent is known as *superplasticity*. Superplasticity is being investigated both for its scientific merit in the context of physical mechanisms operating during plastic deformation and fracture and for its technological significance in the industry. Several review articles summarise both results concerning to superplasticity of various materials and models explaining superplastic flow on the basis of physical mechanisms, e.g., [2–4]. Hexagonal alloys came into the focus of researcher later together with the increasing demand of the industry for magnesium alloys [5, 6].

Three general requirements for the occurrence of superplasticity are necessary:

- Fine microstructure of equiaxed grains with the grain size typically less than 10 μm; reasonably stable during deformation;
- Temperature which is higher than 0.5 T_M (T_M is the melting point temperature in K);
- Strain rate which is typically not too high (higher than $10^{-2}\ s^{-1}$) and not too low (less than $10^{-6}\ s^{-1}$).

Magnesium and magnesium polycrystalline alloys exhibit low plasticity because of their hexagonal structure. For compatible deformation of polycrystals, the activity of five independent slip systems is necessary [7]. Magnesium and magnesium alloys deform in many possible glide systems with dislocations of Burgers vector <***a***> = $1/3\langle 11\bar{2}0\rangle$ in basal, prismatic and first-order pyramidal planes and with dislocations of Burgers vector <***c*** + ***a***> = $1/3<11\bar{2}3>$ in first- and second-order pyramidal planes. The main deformation mode in magnesium is basal slip of <***a***> dislocations in (0001) planes. The secondary conservative slip may be realised by the <***a***> dislocations on prismatic and pyramidal planes of the first order. These slip systems are not crystallographic equivalent with the basal slip system. The critical resolved shear stress for deformation in these systems at lower temperatures is much higher. This shortage of slip systems with easy movable dislocations is the reason for small plasticity of hexagonal metals. Finding the conditions for superplastic deformation may help to overcome this disadvantage. As it will be shown later, superplastic deformation is realised not only by dislocation motion but also by other mechanisms as grain boundary sliding and diffusional transport contributing to the forming of a material. Conditions for the superplastic flow are different for various magnesium alloys depending on their melting point, microstructure, preparation method and diffusional characteristics. The crucial problem is to develop the fine-grained polycrystals with low cost, low forming stress, low forming temperature and high forming rate. In order to reduce grain size, several processes have been used: severe plastic deformation, powder metallurgy techniques, rapid solidification, friction stirring or hot rolling. In this chapter, superplasticity of four commercial magnesium alloys is reported. QE22 and EZ33 alloys contain besides Ag and Zn, respectively, rare earth elements and zirconium. AE42 alloy contains only Al and rare earth elements without Zr. A most frequently used magnesium alloy AZ91, usually prepared as the cast alloy, exhibits very low plasticity. All these alloys exhibit potential of a thermal treatment. Possible mechanisms of the superplastic flow are discussed.

2. Materials and experiments

2.1. AZ91 magnesium alloy

Cast magnesium alloy AZ91 (nominal composition Mg-9Al-1Zn-Mn0.3, in wt%) has a very low ductility especially at low temperatures. A thermal treatment of the alloy may improve the mechanical properties. Formed intermetallics $Al_{12}Mg_{17}$ and Al_8Mn_5 contribute to both high strength and lower plasticity. Various methods were used to achieve grain refinement and to discover the superplastic region.

2.1.1. Hot extrusion

Superplasticity of AZ91 magnesium alloys prepared by hot extrusion (HE) was observed by several authors. Solberg et al. [8] found an excellent superplastic behaviour of rapidly solidified (RS) and extruded alloy. Tensile experiments of a fine-grained AZ91 alloy processed by powder metallurgy (PM) and ingot metallurgy (IM) techniques [9] were carried out at a constant strain rate. The superplastic behaviour was estimated in PM samples at higher strain rates compared with samples prepared with IM route [10]. Results of experiments performed on samples prepared by HE are summarised in **Table 1**.

Preparation route	Grain size (μm)	*T* (°C)	SR (s^{-1})	*m*	ε_f (%)	Remark	References
RS + HE	1.2	300	2.5×10^{-3}	0.60	1480		[8]
PM + HE	1.4	300	1×10^{-2}	0.50	280*	$T/T_M = 0.62$	[9]
PM + HE	4.1	250	3×10^{-3}	0.50	430*	$T/T_M = 0.57$	[9]
IM + HE	5	300	3×10^{-3}	<0.3	425*		[10]

*Tensile test at constant strain rate.

Table 1. Superplastic characteristics and conditions estimated for AZ91 alloy prepared by hot extrusion.

2.1.2. Thermomechanical treatment

An alternative route to prepare superplastic alloy is the thermomechanical treatment (TMT) consisted from the heat treatment in two stages (homogenisation at 470°C for 10 h and ageing at temperatures from 200 to 380°C). Aged samples were hot extruded at 350°C. [11–13]. The grain size of samples and superplastic behaviour depends on the ageing temperature. This method is relatively simple and big bulks of the material may be prepared. It is an advantage against other methods, for example, high-pressure torsion technique, which is able to manufacture only small samples. Results of the superplastic behaviour of AZ91 alloys prepared by the TMT are summarised in **Table 2**.

Preparation route	Grain size (μm)	T (°C)	SR (s^{-1})	m	ε_f (%)	Remark	References
TMT	17	420	1×10^{-4}	0.44	413*	Aged 200°C	[11]
TMT	17	380	3×10^{-4}		285	Aged 200 °C	[12]
TMT	10.5	420	1×10^{-4}	0.54	484*	Aged 300°C	[11]
TMT	10.5	380	3×10^{-4}		280	Aged 300°C	[12]
TMT	10.5	340	3×10^{-4}		291	Aged 300°C	[12]
TMT	11	420	1×10^{-4}	0.54	584*	Aged 380°C	[11]
TMT	11	380	3×10^{-4}		251	Aged 380°C	[12]
TMT	11	340	3×10^{-4}		260	Aged 380°C	[12]
TMT	6.4	420	1×10^{-4}	0.45	250*	Aged 350°C	[13]
TMT	6.4	420	3×10^{-4}	0.40	260*	Aged 350°C	[13]
TMT	11.2	400	3×10^{-4}	0.35	250	Aged 300°C	[14]

*Deformed at a constant strain rate.

Table 2. Superplastic characteristics and conditions estimated for AZ91 alloy prepared by thermomechanical treatment.

2.1.3. *Equal channel angular pressing*

Excellent superplasticity of samples prepared by equal channel angular pressing (ECAP) was observed by Chuvil'deev et al. [15] at a relatively high temperature of 300 °C. Mussi et al. [16] found the superplastic behaviour after ECAP at a constant temperature (eight passes) of 265°C (CT) and at decreasing temperature in each following pass from 265 down to 150°C in the last one. A substantial grain refinement was observed for both ECA extrusion techniques. ECAP technique with decreasing temperature caused a decrease in the grain size up to sub-microcrystalline region. Matsubara et al. [17] used (first time) a new method for preparation of fine-grained materials involving sequential application of hot extrusion and ECAP. This method is in the literature designated as EX-ECAP. Mabuchi et al. [18] determined the superplastic behaviour of an AZ91 alloy after hot extrusion and ECAP. Very small grain size (about 1 μm) was very probably the reason for the occurrence of superplastic flow at a relatively low temperature of $0.48T_M$. Superplasticity of fine-grained AZ91 alloy prepared by ECAP technique and annealing was studied by Mabuchi and co-workers [19]. Subsequent annealing for 12 h at 225°C improved the superplastic behaviour of the alloy although the grain size increased 4.4 times. The authors explained this finding to the transformation of non-equilibrium grain boundaries in the as ECAPed material to equilibrium grain boundaries which were detected in the annealed alloy by means of the transmission electron microscopy. The superplastic behaviour of AZ91 alloys prepared by the hot extrusion and ECAP or by a combination of both techniques is summarised in **Table 3**.

Preparation route	Grain size (μm)	*T* (°C)	SR (s^{-1})	*m*	ε_f (%)	Remark	References
ECAP-6p.	0.8	300	3×10^{-3}	0.3	570	Q = 78 kJ/mol	[15]
ECAP-6p.	0.8	250	3×10^{-2}	0.3	375	Route B_C	[15]
ECAP-8p.	1.0	300	2×10^{-4}		340	CT	[16]
ECAP-8p.	1.0	250	1×10^{-3}		190	CT	[16]
ECAP-8p.	0.3	250	1×10^{-3}		490	DT	[16]
HE-ECAP	1.0	200	6×10^{-5}	0.30	661	$T/T_M = 0.51$	[18]
HE-ECAP	1.0	175	6×10^{-5}	0.30	326	$T/T_M = 0.48$	[18]
ECAP	0.7	200	6×10^{-5}	0.5	650	Q = 89 kJ/mol	[19]
ECAP	0.7	150	6×10^{-5}	0.5	620	Non-eq. GB	[19]
ECAP + ann.*	3.1	200	7×10^{-5}	0.5	990	Q = 89 kJ/mol	[19]
ECAP + ann.*	3.1	250	8×10^{-5}	0.5	980	Eq. GB	[19]

*ECAP with subsequent annealing.
Non-eq. GB: non-equilibrium grain boundaries; Eq. GB: equilibrium grain boundaries.

Table 3. Superplastic characteristics and conditions estimated for AZ91 alloy prepared by the ECAP technique.

2.1.4. Hot rolling

Wei et al. [20] prepared AZ91 sheet by repeated rolling (RR) at 400°C. The rolls were annealed between rolling passes at 130°C. An initial grain size of 78 μm decreased after rolling down to 11 μm. Resulting grains were equiaxed and no twins were observed in the microstructure. Rolling and heating were applied 11 times. Samples for tensile deformation at elevated temperature have the tensile axis parallel to the rolling direction. Interesting results were obtained by Mohri et al. [21] on hot-rolled (HR) sheets. High-ratio differential speed rolling (HRDSR) was used to refine the microstructure of the cast alloy [22]. Very fine particles of the γ-phase hinder the grain growth during the high-temperature straining. An increase in temperature from 300 to 350°C reduced substantially the plasticity. The superplasticity characteristics of samples prepared by hot rolling are given in **Table 4**.

Preparation route	Grain size (μm)	*T* (°C)	SR (s^{-1})	*m*	ε_f (%)	Remark	References
RR (400°C)	11	350	1.5×10^{-3}	0.64	455	*Q* = 92.7 kJ/mol	[20]
HR (400°C)	39.5	300	1.5×10^{-3}		601	sub-grains~1μ	[21]
HRDSR	1–2	300	5.0×10^{-4}		550		[22]
HRDSR	1–2	300	1.0×10^{-3}	0.5	830		[22]
HRDSR	1–2	300	1.0×10^{-2}		450		[22]
HRDSR	1–2	350	1.0×10^{-3}		250		[22]

Table 4. Superplastic characteristics and conditions estimated for AZ91 alloy prepared by the hot rolling.

2.1.5. Friction stirring

Friction stir processing (FSP) is usually used for the friction stir welding purposes. Mishra and Ma [23] showed that this method may be applied as a novel grain refinement technique. In this case, the stirring action of the tool on the bulk material is used for a forging/extrusion treatment at very high temperatures. Dynamic recrystallization and grain rotation leads to a very fine microstructure. The superplastic behaviour of AZ91 magnesium alloy and on FS-processed AZ91 produced by high-pressure die casting was observed by Cavaliere and Marco [24]. Raja and co-workers [25] estimated that the microstructure depends sensitively on the FS conditions. The highest elongation to fracture was found for samples processed on one side up to a depth of 5 mm. An excellent superplasticity behaviour was observed by Zhang et al. [26]. They estimated very fine γ-$Mg_{17}Al_{12}$ and Al-Mn intermetallics which effectively hinder the grain growth at elevated temperatures. The variation of this method consists in some additional cooling during FSP. Submerged friction stir processing (SFSP) is realised underwater and it has a great potential in the preparation of ultrafine-grained materials [27]. Chai et al. [28] by applying the SFSP on AZ91 alloy obtained a substantial refinement of the grain structure and a considerably enhanced superplastic behaviour. Elongations to fracture of samples prepared with the conventional FSP and SFSP are compared in **Table 5**. The grain size of samples prepared with the SFSP decreased to 1.2 μm compared with FSP performed on air. An excellent superplastic behaviour of SFSP samples especially at higher strain rates was ascribed to finer structure and larger fraction of grain boundary [28]. A combination of the high-pressure die casting (HPDC) with the FSP resulted into ultrafine material with the grain size lower than 1 μm and excellent superplastic properties at 330°C and at relatively high strain rates [29]. Superplastic behaviour of AZ91 alloys prepared with routes exploited FSP is summarised in **Table 5**.

Preparation route	Grain size (μm)	T (°C)	SR (s^{-1})	m	ε_f (%)	Activation energy (kJ/mol)	References
FSP		300	5×10^{-4}	0.68	1050		[24]
FSP		350	5×10^{-4}	0.6	680		[25]
FSP	3	300	3×10^{-3}		240	175.05	[26]
FSP	3	300	4×10^{-4}		517	175.05	[26]
FSP	3	300	1×10^{-4}		1604	175.05	[26]
FSP	7.8	350	4×10^{-4}	0.36	554	125.91 (148.75)*	[28]
FSP	7.8	350	3×10^{-3}	0.36	402	125.91 (148.75)*	[28]
FSP	7.8	350	2×10^{-2}	0.25	158		[28]
SFSP	1.2	350	4×10^{-4}	0.43	990	88.30 (168.12)*	[28]
SFSP	1.2	350	3×10^{-3}	0.43	1202	88.30 (168.12)*	[28]
SFSP	1.2	350	2×10^{-2}	0.30	990		[28]
HPDC + FS	0.5	330	1×10^{-2}		1251		[29]
HPDC + FS	0.5	330	3×10^{-2}		827		[29]

*Activation energy was estimated for the temperature ranges of 200–350°C (lower value) and 350–375°C (higher value).

Table 5. Superplastic characteristics and conditions estimated for AZ91 alloy prepared by the friction stir processing.

2.1.6. *High-pressure torsion*

Samples processed by high-pressure torsion (HPT), (10 turns), have a fine-equiaxed grain structure with a significant thermal stability [30]. The thermal stability was due to high volume fraction of nanosized γ-phase particles. The main mechanism is grain boundary sliding accommodated by dislocation creep process and diffusional process. The fine fibrous microstructure resisted cavitation and linked the disconnected grains [30]. The experimental results concerning to the superplastic behaviour of AZ31 alloy prepared by HPT are given in **Table 6**.

Preparation route	T (°C)	SR (s^{-1})	m	ε_f (%)	References	Remark
HPT	200	1×10^{-2}	0.42	590	[30]	
HPT	200	1×10^{-3}	0.46	660	[30]	
HPT	200	1×10^{-4}	0.46	670	[30]	
HPT	300	1×10^{-1}	0.41	410	[30]	
HPT	300	1×10^{-2}	0.42	860	[30]	
HPT	300	1×10^{-3}	0.43	1050	[30]	10 turns
HPT	300	1×10^{-4}	0.52	1308	[30]	10 turns

Table 6. Superplastic characteristics and conditions estimated for AZ91 alloy prepared by high-pressure torsion.

2.2. QE22 alloy

Mg-2Ag-2RE-0.7Zr alloys were developed for applications at elevated temperatures. Further improvement of mechanical properties is possible using some heat treatments producing thermally stable precipitates [31, 32].

2.2.1. *Thermomechanical treatment*

Thermomechanical processing for the production of fine-grained materials used by several authors [13, 14, 33] consists of subsequent steps: solution treatment at 470°C for 10 h quenched on air; and aged for 10 h followed by hot extrusion at a temperature of 350°C. Three ageing temperatures were applied with the aim to find optimum conditions for the superplastic deformation. Microstructure of the alloy consists from α-grains decorated in grain boundaries by the chain of intermetallic particles containing Nd and Ag. Tiny Zr particles (100–200 nm) were not dissolved and affected by the thermal treatment. The activation energy was calculated to be Q = 114.8 kJ/mol [34]. **Table 7** summarises superplastic characteristics.

Preparation route	Grain size(μm)	T (°C)	SR (s^{-1})	m	ε_f (%)	Remark	References
TMT	6.1	420	2×10^{-4}	0.65	490	Ageing 380°C	[34]
TMT	1.0	450	3.3×10^{-4}	0.71	880*	Ageing 350°C	[33]
TMT	1.0	480	3.3×10^{-4}	0.66	910*	Ageing 300°C	[33]
TMT	0.7	420	1×10^{-4}	0.62	720	Ageing 300°C	[13]

Preparation route	Grain size(μm)	T (°C)	SR (s^{-1})	m	ε_f (%)	Remark	References
TMT	0.7	420	3×10^{-4}	0.75	780	Ageing 300°C	[13]
TMT	0.7	420	5×10^{-4}	0.50	450	Ageing 300°C	[13]
TMT	0.7	420	1×10^{-3}	0.42	360	Ageing 300°C	[13]
TMT	0.7	420	5×10^{-3}	0.38	300	Ageing 300°C	[13]
TMT	0.7	420	0.01	0.32	240	Ageing 300°C	[13]
TMT	1.9	400	3×10^{-4}	0.75	750*	Ageing 380°C	[14]

*Straining at a constant crosshead speed.
Tensile tests were performed at a constant strain rate.

Table 7. Superplastic characteristics and conditions estimated for QE22 alloy prepared by TMT.

2.2.2. Friction stir process

Samples for tensile tests were friction stir processed in a temperature-controlled atmosphere [35]. In order to increase the efficiency of the FS, a device external cooling set-up was used. The FSP was performed in two passes with different tool rotation rates. Refinement of the microstructure was really significant from 38 μm down to 630 nm (**Table 8**). Conditions for superplasticity and elongation to fracture of samples are reported in **Table 8**.

Preparation route	Grain size (μm)	T (°C)	SR (s^{-1})	m	ε_f (%)	Remark	References
FSP	0.63	450	1×10^{2}		1630	Q = 170 kJ/mol	[35]
FSP	0.63	425	1×10^{2}		~1000		[35]
FSP	0.63	350	3×10^{3}	0.5	850		[35]
FSP	0.63	350	1×10^{2}	0.5	450		[35]
FSP	0.63	300	5×10^{-4}		420	Q = 142 kJ/mol	[35]

Table 8. Superplastic characteristics and conditions estimated for QE22 alloy prepared by FSP.

2.3. AE42 alloy

In AE42 alloy, fine Mg_9RE precipitates along the grain boundaries are formed. Aluminium improves castability and room temperature mechanical properties [36].

The AE42 magnesium alloy (Mg-4Al-2RE) casting was subjected to the subsequent two-stage thermal treatment: homogenisation (10 h at 470°C) and ageing at 300 °C and 350 °C, followed hot extrusion at 350 °C. Thermomechanical treatment refined the microstructure under 10 μm [13] and increased plasticity of samples [13, 14] as it is shown in **Table 9**.

Preparation route	Grain size (μm)	T (°C)	SR (s^{-1})	m	ε_f (%)	Remark	References
TMT	6.2	420	1×10^{-4}	0.43	235	Ageing 300°C	[13]
TMT	6.2	420	3×10^{-4}	0.37	220	Ageing 300°C	[13]
TMT	6.2	420	5×10^{-4}	0.36	180	Ageing 300°C	[13]
TMT	6.2	420	1×10^{-3}	0.30	150	Ageing 300°C	[13]
TMT	13.5	400	1×10^{-4}	0.4	220	Ageing 350°C	[14]

Tensile tests were performed at a constant strain rate.

Table 9. Superplastic characteristics and conditions estimated for AE42 alloy prepared by TMT.

2.4. EZ33 alloy

The Mg-RE-Zn alloys have additional rare earth elements to improve the creep resistance and to refine the grain size. Further increase of strength occurs if zinc is added [37]. The continuous networks of intergranular phases in Mg-Zn-Nd-Zr alloys significantly deteriorate the ultimate tensile strength and elongation [38] Superplastic samples of the EZ33 (Mg-2.5Zn-3RE-Zr) magnesium alloy were prepared with the same thermomechanical treatment as AZ91, QE22 and AE42 alloys. Precipitation ageing was realised at 470°C for 10 h with a subsequent cooling on the air. As it follows from **Table 10**, the tensile elongations to fraction for all samples exceeded 200%, which represents a substantial improvement over the poor room-temperature ductility; typical for magnesium alloys [13, 34].

Similar alloy Mg-3Gd-1Zn (GZ31) was processed by EX-ECAP and deformed by shear punch testing [39]. Samples with very fine microstructure (grain size 1.7 μm) were deformed in shear punch tests in the temperature interval from 300 to 500°C. A strain rate parameter of 0.51 and activation energy of 73 kJ/mol indicate superplastic behaviour — see **Table 10**.

Preparation route	Grain size (μm)	T (°C)	SR (s^{-1})	m	ε_f (%)	Remark	References
TMT	1.2	420	1×10^{-4}	0.65	700	Ageing 350°C	[13]
TMT	1.2	420	3×10^{-4}	0.73	710	Ageing 350°C	[13]
TMT	1.2	420	5×10^{-4}	0.50	420	Ageing 350°C	[13]
TMT	1.2	420	1×10^{-3}	0.48	340	Ageing 350°C	[13]
TMT	1.2	420	5×10^{-3}	0.35	280	Ageing 350°C	[13]
TMT	1.2	420	0.01	0.30	230	Ageing 350°C	[13]
TMT	2.2	420	4×10^{-4}	0.36	473	Ageing 300°C	[34]
TMT	2.2	420	2×10^{-4}	0.38	490	Ageing 300°C	[34]
EX-ECAP	1.7	400	1×10^{-3}	0.51		GZ31 73 kJ/mol	[39]

Tensile tests were performed at a constant strain rate

Table 10. Superplastic characteristics and conditions estimated for EZ33 alloy prepared by TMT.

3. Discussion

3.1. Grain refinement

Results obtained for various materials prove that the grain size has the key role for the occurrence of superplastic flow. It follows from relationship (3), (the inverse dependence on grain size in a power of 2). The second necessary condition which should be fulfilled to achieve superplastic deformation is the thermal stability of grains.

If superplastic materials are prepared by some thermomechanical treatment, the processing route should introduce (produce) a number of recrystallization nuclei in the alloys through homogenisation, ageing with a successive hot extrusion process. Comparing resulting grain sizes of samples prepared by the TMT, it is to see that the grains in the alloys containing Zr are much finer. The finer grains in the magnesium alloys QE22 and EZ33 alloys may be attributed to small Zr particles and RE-rich precipitates formed in both materials during the thermal treatment.

It should be mentioned that the grain refinement in hexagonal alloys is different from that in fcc metals [40]. The grain refinement in hexagonal alloys subjected to ECAP starts by the nucleation of tiny grains in pre-existing grain boundaries [41, 42]. The stress concentration in the grain boundaries may activate non-basal slip. A co-operation of basal and non-basal slip is the prerequisite for formation of three-dimensional net of equiaxed grains. Reducing number of passes and increasing the processing temperature may shift the superplasticity region to higher temperatures [43].

Fragmentation of second-phase particles was observed in QE22 alloy after multi-pass FSP [44]. Very high plastic deformation is introduced into the stirred zone during FSP. Disintegrated particles serve as nuclei in the dynamic recrystallization. At high strain rate superplasticity, a rapid growth of cavities is due to a significant stored plastic deformation and lattice diffusion. Similar mechanisms of grain refinement may be considered in the high-pressure torsion producing: disintegration of intermetallics followed by dynamic recrystallization (DXR) observed by Száraz et al. [45] in Mg-based nanocomposites subjected to high-pressure torsion.

3.2. Mechanisms of superplastic flow

Superplastic flow of metallic materials may be described using constitutive equation [46]:

$$\dot{\varepsilon} = A\left(\frac{Gb}{kT}\right)\left(\frac{\sigma - \sigma_0}{G}\right)^n\left(\frac{b}{d}\right)^p D, \tag{3}$$

where A is a dimensionless material constant, G is the shear modulus, b is the Burgers vector, σ is the applied stress, σ_0 is the threshold stress, d is the grain size, n is the stress sensitivity exponent (n = 1/m), p is the grain size exponent and kT has its usual meaning. D is the diffusion coefficient (=$D_0 \exp(-Q/RT)$, where D_0 is the frequency factor, Q is the activation energy for the diffusion process and R is the gas constant).

Plastic deformation of coarse-grained materials at temperatures $T < 0.4T_M$ is realised by dislocation motion and storage of dislocations on obstacles. These obstacles may be of two types: dislocation and non-dislocation type (grain boundaries, twins′ boundaries and incoherent precipitates). Storage of dislocations contributes to strain hardening of a material. On the other hand, recovery processes like cross-slip and dislocation climbing play an important role especially at elevated temperatures and contribute to softening. Steady-state character of the stress-strain curves is a consequence of a dynamic equilibrium between hardening and softening processes. In fine-grained materials, two additional deformation mechanisms—grain boundary sliding accommodated by slip and diffusion flow—should be considered [46, 47].

Steady-state plastic flow of coarse-grained metals at higher temperatures, above $0.4T_M$, is usually described by dislocation motion and storage on obstacles. Each mechanism—dislocation slip, grain boundary sliding and diffusional flow—has its characteristic values of stress sensitivity, n, grain size exponent, p and the activation energy, Q. Plastic materials have typically high values of the stress sensitivity parameter $n = 1/m$. For $n \geq 5$, we may consider that the main mechanism of plastic deformation is the dislocation motion. The activation energy is done by the strength of local obstacles in the thermally activated dislocation motion. If this dislocation motion is non-conservative—climb of dislocations (for example, in the case of the high-temperature creep)—the characteristic activation energy is close to the activation energy for the lattice diffusion or pipe diffusion. The grain size exponent has very low value approaching zero [47]. As it was shown above, high values of the strain rate sensitivity parameter (in the case of ideal superplasticity $m = 1$) are characteristic for the superplastic flow, i.e., the stress sensitivity parameter must be from the interval $1 < n < 3$. The apparent activation energy may help to identify the deformation mechanism(s) occurring in the superplastic flow [47].

$$Q = \frac{1}{m} R \frac{\Delta ln\,(\sigma/G)}{\Delta(1/T)}, \tag{4}$$

where σ is the applied stress, R is the gas constant and T is the absolute temperature. The activation energy estimated for lattice and grain boundary diffusion in magnesium is 135 and 92 kJ/mol, respectively [48]. The activation energies reported in **Tables 3–5, 8** and **10** have values which are between both values estimated for the lattice and grain boundary diffusion. It indicates that very probably both mechanisms are present. Activation energies introduced in **Table 8** for QE22 alloy are higher than the values measured for lattice diffusion. Arzt and co-workers [49] have shown that the overcoming of intergranular precipitates by grain boundary dislocations leads to a higher activation energy [49].

It is interesting to note that mechanical twinning as an additional deformation mechanism was observed in some cases.

Figure 1 shows scanning electron micrographs (SEM micrographs) of AZ91 and QE22 samples deformed in the superplastic region. Deformation twins are visible in both micrographs 1a and 1b. They provide evidence that in Mg alloys, twinning may serve as an additional accommodation mechanism for grain boundary sliding at straining temperatures which do not allow for

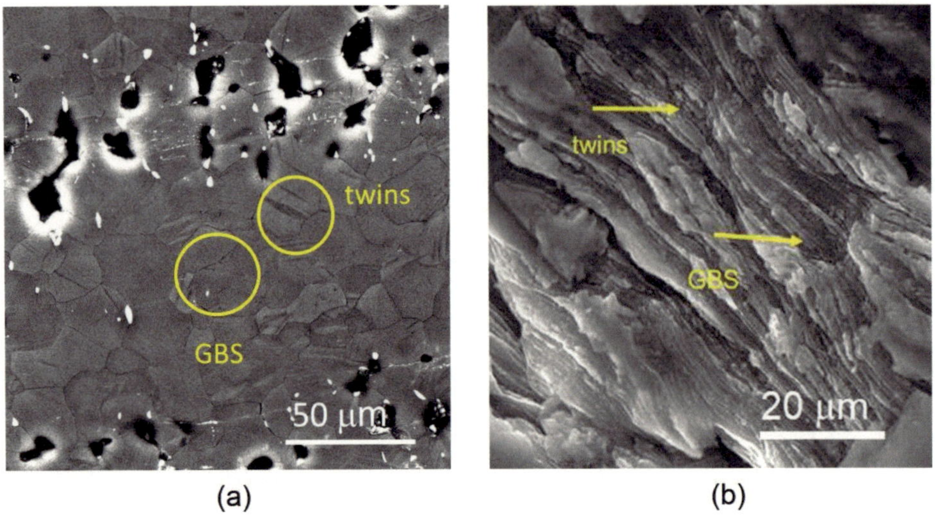

Figure 1. (a) Twins and grain boundary sliding in the AZ91sample deformed at 340 °C and (b) surface wrinkles estimated after deformation of QE22 alloy at 450°C.

sufficiently fast diffusion-related accommodation. On the other hand, transmission electron microscopy revealed in QE22 and EZ33 alloys clear grains, without dislocation in deformed samples at 420°C as it has been reported in our previous paper [50].

3.3. Cavitation during superplastic flow

Cavitation in the AZ91 alloy occurred during superplastic deformation at 420°C. Similarly, chains of cavities are obvious in the micrograph of the AZ91 alloy deformed at 340°C.

The formation and growth of cavities relaxes the stress concentration in triple junctions caused by grain boundary sliding (GBS). Stress concentrations are likely developed at sites which impede grain boundary sliding such as grain boundary particles, ledges and triple points. Cavities created by vacancy clustering may nucleate if the stress concentration is not relieved sufficiently rapidly. The AZ91 alloy with intermetallic inclusions can be considered as a natural composite. The stress concentrations are formed at the particles on sliding grain boundaries. Local tensile stress caused by sliding at interfaces may be expressed in the following form [19]:

$$\sigma_{slid} = \frac{0.92kTd_p\dot{\varepsilon}dV_f}{\Omega D_L\left(1+5\frac{\delta D_{GB}}{d_pD_L}\right)} \tag{5}$$

where d_p is the particle diameter, $\dot{\varepsilon}$ is the strain rate, D_L is the lattice diffusion coefficient and D_{GB} is the grain boundary diffusion coefficient, δ is the grain boundary width, Ω is the atomic volume and V_f is the volume fraction of particles. Other symbols have the same meaning as in Eq. (3). Mechanical twinning observed in AZ91 alloy deformed at 340°C and in QE22 alloy deformed at 430°C may serve as an additional accommodation mechanism for grain boundary sliding at straining temperatures which do not allow sufficiently fast diffusion-related accommodation. The insufficiently accommodated GBS process is documented in **Figures 1a** and **2**, where cavities formed during deformation of AZ91 alloy are visible. Chains of cracks are oriented in the extrusion direction; they show that this process is also influenced by the basal texture of materials formed during the hot extrusion. The cracks were developed from small voids created as a result of the grain boundary sliding. The influence of texture is also visible in **Figure 1b**, where observed wrinkles are traces of basal planes which were oriented, in the textured sample, parallel to the extrusion direction and sample axis.

The presented results allow us to conclude that grain boundary sliding is very probably the main deformation mechanism during the superplastic deformation in the reported alloys. The different thermomechanical treatment influenced the samples microstructure and also different contributions to the superplastic flow: grain boundary sliding, lattice/grain boundary diffusion, dislocation deformation (slip of basal and non-basal dislocations, creep and climb of dislocations) and mechanical twinning.

Figure 2. Cavities in AZ91 sample deformed at 420 °C and constant strain rate of 1×10^{-4} s^{-1}. Arrow indicates tension direction.

4. Concluding remarks

The superplastic behaviour of four magnesium alloys has been reported. Occurrence of superplasticity requires a special microstructure containing small equiaxed grains and stable grain boundaries. Various methods were used for preparation of materials corresponding to these requirements. Thermomechanical treatment is the method which is relatively simple and gives acceptable results. Bulks of materials may be processed and used for possible applications in the industry. On the other hand, methods, exploited severe plastic deformation or friction stir processing, are still an academic issue. The main problem of superplasticity is the transfer of results achieved in the laboratory into the industry. Applications of magnesium alloys in various industrial branches continuously increase. Owing to small plasticity of these alloys, the finding of the conditions for superplastic flow plays a very important role.

Acknowledgements

This chapter is dedicated to the memory of Oscar Akramovich Kaibyshev founder and former director of Institute of Metals Superplasticity Problems of the Russian Academy of Sciences, who has contributed very much to a better understanding of superplastic behaviour of various materials.

Authors gratefully acknowledge the financial support from the Charles University under grant PROGRES Q47.

Conflict of interest

Authors declare no conflict of interest.

Author details

Zuzanka Trojanová*, Zdeněk Drozd and Pavel Lukáč

*Address all correspondence to: ztrojan@met.mff.cuni.cz

Faculty of Mathematics and Physics, Charles University, Prague, Czech Republic

References

[1] Reid CN. Deformation Geometry for Scientists. 1st ed. Oxford: Pergamon Press; 1973. p. 220. DOI: 10.1016/C2013-0-02546-7

[2] Mukherjee AK. Superplasticity in metals, ceramics and Intermetallics. In: Cahn RW, Haasen P, Kramer EJ, editors. Materials Science and Technology Vol. 9.: Deformation and Fracture of Materials (Volume Editor Mughrabi H). Weinheim: Wiley-VCH; 2006. pp. 408-456. DOI: 10.1002/9783527603978.mst0056

[3] Chokshi AH, Mukherjee AK, Langdon TG. Superplasticity in advanced materials. Materials Science and Engineering R. 1993;**10**:237-274. DOI: 10.1016/0927-796X(93)90009-R

[4] Figueiredo RB, Sabbagahianard S, Giwa A, Greer J, Langdon TG. Evidence for exceptional low temperature ductility in polycrystalline magnesium processed by severe plastic deformation. Acta Materialia. 2017;**122**:322-331. DOI: 10.1016/j.actamat.2016.09.054

[5] Watanabe H, Mukai T, Mabuchi M, Higashi K. Superplastic deformation mechanisms in powder metallurgy magnesium alloys and composites. Acta Materialia. 2001;**49**:2027-2037. DOI: 10.1016/S1359-6454(01)00101-X

[6] Watanabe H, Mukai T, Ishikawa K, Mabuchi M, Higashi K. Realization of high-rate superelasticity at low temperatures in a Mg-Zn-Zr alloy. Materials Science and Engineering A. 2001;**307**:119-128. DOI: 10.1016/S0921-5093(00)01974-2

[7] Mises RV. Mechanik der plastichen Formänderung von Kristallen. Zeitschrift für Angewandte Mathematik und Mechanik. 1928;**8**:161-185. DOI: 10.1002/zamm.19280080302

[8] Solberg JK, Tørklep J, Bauger Ø, Gjestland H. Superplasticity in magnesium alloy AZ91. Materials Science and Engineering A. 1991;**134**:1201-1203. DOI: 10.1016./0921-5093(91)90955-M

[9] Mabuchi M. Experimental investigation of superplastic behaviour in magnesium alloys. Materials Science and Technology. 1997;**13**:825-831. DOI: 10.1179/mst.1997.13.10.825

[10] Kubota K, Mabuchi M, Higashi K. Review processing and mechanical properties of fine-grained magnesium alloys. Journal of Materials Science. 1999;**34**:2255-2262. DOI: 10.1023/A:1004561205627

[11] Trojanová Z, Száraz Z, Ryspaev T, Wesling V. Influence of thermomechanical treatment on the superplastic behaviour of an AZ91 magnesium alloy. Kovove Materialy. 2008;**46**:285-289

[12] Trojanová Z, Turba K, Száraz Z, Ryspaev T, Palček P, Chalupová M. Superplastic behaviour of an extruded AZ91 alloy. Acta Physica Polonica, A. 2012;**122**:597-600. DOI: 10.12693/APhysPolA.122.597

[13] Wesling V, Ryspaev T, Schram A. Effect of overageing temperature on the superplastic behaviour in magnesium alloys. Materials Science and Engineering A. 2007;**462**:144-148. DOI: 10.1016/j.msea.2006.03.152

[14] Ryspaev T, Janeček M, Král R, Weling V, Wagner L. Processing, superplastic properties and friction stir welding of fine-grained AZ31, AZ91, AE42 and QE22 magnesium alloys. Materials Science Forum. 2016;**838-839**:220-224. DOI: 10.4028/www.scientific.net/MSF.838-839.220

[15] Chuvil'deev VN, Nieh TG, Gryaznov MY, Kopylov VI, Sysoev AN. Superplasticity and internal friction in microcrystalline AZ91 and ZK60 magnesium alloys processed by equal-channel angular pressing. Journal of Alloys and Compounds. 2004;**378**:253-257. DOI: 10.1016/j.jallcom.2003.10.062

[16] Mussi A, Blandin JJ, Rauch EF. Microstructure refinement and improvement of mechanical properties of a magnesium alloy be severe plastic deformation. In: Proceedings of the International Conference on Nanomaterials by Severe Plastic Deformation (NANOSPD2); 9–13 December 2002; Vienna. Weinheim: Wiley-VCH; 2004. pp. 740-745. DOI: 10.1002/3527602461.ch13g

[17] Matsubara K, Miyahara Y, Horita Z, Langdon TG. Developing superplasticity in a magnesium alloy through a combination of extrusion and ECAP. Acta Materialia. 2003;**51**:3073-3084. DOI: 10.1016/S1359-6454(03)00118-6

[18] Mabuchi M, Iwasaki H, Yanasa K, Higashi K. Low temperature superplasticity in an AZ91 magnesium alloy processed by ECAE. Scripta Materialia. 1997;**36**:681-686. DOI: 10.1016/S1359-6462(96)00444-7

[19] Mabuchi M, Ameyamab K, Iwasaki H, Higashi K. Low temperature superplasticity of AZ91 magnesium alloy with non-equilibrium grain boundaries. Acta Materialia. 1999;**47**: 2047-2057. DOI: 10.1016/S1359-6454(99)00094-4

[20] Wei YH, Wang QD, Zhu YP, Zhou HT, Ding WJ, Chino Y, Mabuchi M. Superplasticity and grain boundary sliding in rolled AZ91 magnesium alloy at high strain rates. Materials Science and Engineering A. 2003;**360**:107-115. DOI: 10.1016/S0921-5093(03)00407-6

[21] Mohri T, Mabuchi M, Nakamura M, Asahina T, Iwasaki H, Aizawa T, Higashi K. Microstructural evolution and superplasticity of rolled Mg-9Al-1Zn. Materials Science and Engineering A. 2000;**290**:139-144. DOI: 10.1016/S0921-5093(00)00959-X

[22] Kim YS, Kim WJ. Microstructure and superplasticity of the as-cast Mg-9Al-1Zn magnesium alloy after high ratio differential speed rolling. Materials Science and Engineering A. 2016;**677**:332-339. DOI: 10.1016/j.msea.2016.09.063

[23] Mishra RS, Ma ZY. Friction stir welding and processing. Materials Science and Engineering R. 2005;**50**:1-78. DOI: 10.1016/j.mser.2005.07.001

[24] Cavaliere P, De Marco PP. Superplastic behavior of friction stir processed AZ91 magnesium alloy produced by high pressure die cast. Journal of Materials Processing Technology. 2007;**184**:77-83. DOI: 10.1016/j.jmatprotec.2006.11.005

[25] Raja A, Biswas P, Pancholi V. Effect of layered microstructure on the superplasticity of friction stir processed AZ91 magnesium alloy. Materials Science and Engineering A. 2018; **725**:492-502. DOI: 10.1016/j.msea.2018.04.028

[26] Zhang D, Wang S, Qiu C, Zhang W. Superplastic tensile behavior of a fine grained AZ91 magnesium alloy prepared by friction stir processing. Materials Science and Engineering A. 2012;**556**:100-106. DOI: 10.1016/j.msea.2012.06.063

[27] Hofmann DC, Wecchio KS. Submerged friction stir processing (SFSP): An improved method for creating ultra-fine-grained bulk material. Materials Science and Engineering A. 2005;**402**(1–2):234-241. DOI: 10.1016/j.msea.2005.04.032

[28] Chai F, Zhang D, Li Y, Zhang W. High strain rate superplasticity of a fine-grained AZ91 magnesium alloy prepared by submerged friction stir processing. Materials Science and Engineering A. 2013;**568**:40-48. DOI: 10.1016/j.msea.2013.01.026

[29] Mohan A, Yuan W, Mishra RS. High strain rate superplasticity in friction stir processed ultrafine-grained Mg-Al-Zn alloys. Materials Science and Engineering A. 2013;**562**:69-76. DOI: 10.1016/j.msea.2012.11.026

[30] Al-Zubaydi ASJ, Zhilyaev AP, Wang SC, Reed PAS. Superplastic behaviour of AZ91 magnesium alloy processed by high-pressure torsion. Materials Science and Engineering A. 2015;**637**:1-11. DOI: 10.1016/j.msea.2015.04.004

[31] Rokhlin LL. Magnesium Alloys Containing Rare Earth Metals: Structure and Properties. London: CRC Press; 2003. p. 256. ISBN: 9780415284141

[32] Khan MDF, Panigrahi SK. Age hardening, fracture behavior and mechanical properties of QE22 Mg alloy. Journal of Magnesium and Alloys. 2015;**3**:210-217. DOI: 10.1016/j.jma.2015.08.002

[33] Trojanová Z, Chiappa S, Málek P, Száraz Z, Lukáč P, Ryspaev T. Superplastic behaviour of an Mg-Ag-RE magnesium alloy. Acta Physica Polonica, A. 2015;**128**:765-767. DOI: 10.12693/APhysPolA.128.765

[34] Trojanová Z, Szaraz Z, Padalka O, Ryspaev T, Lukáč P. Structural (super)plasticity of magnesium materials. Komunikacie (Communications). 2012;**14**:19-25. ISSN: 13354205

[35] Khan MDF, Panigrahi SK. Achieving excellent superplasticity in an ultrafine grained QE22 alloy at both high strain rate and low temperature regime. Journal of Alloys and Compounds. 2018;**747**:71-82. DOI: 10.1016/j.jallcom.2018.02.294

[36] Wei LY, Dunlop GL. Precipitation hardening in a cast Mg-rare earth alloy. In: Proceedings of the International Conference on Magnesium Alloys and their Applications; 17 December 1992. Germany: DGM Informationsgesellschaft; 1992. pp. 335-342

[37] Ben-Hamu G, Eliezer D, Shin KS, Cohen S. The relation between microstructure and corrosion behavior of Mg-Y-Zr alloys. Journal of Alloys and Compounds. 2007;**431**(1–2): 269-276. DOI: 10.1016/j.jallcom.2006.05.075

[38] Li Q, Wang Q, Wang Y, Zeng X, Ding W. Effect of Nd and Y additions on microstructure and mechanical properties of as-cast Mg-Zn-Zr alloy. Journal of Alloys and Compounds. 2007;**427**:115-123. DOI: 10.1016/j.jallcom.2006.02.054

[39] Sarebanzadeh M, Roumina R, Mahmudi R, Wu GH, Jafari Nodooshan HR. Enhancement of superplasticity in a fine-grained Mg-3Gd-1Zn alloy processed by equal-channel angular pressing. Materials Science and Engineering A. 2015;**646**:249-253. DOI: 10.1016/j.msea.2015.08.074

[40] Mishra RS, Bieler TR, Mukherjee AK. Mechanism of high strain rate superplasticity in aluminium alloy composites. Acta Materialia. 1997;**45**:561-568. DOI: 10.1016/S1359-6454(96)00194-2

[41] Figueiredo RB, Langdon TG. Principles of grain refinement and superplastic flow in magnesium alloys processed by ECAP. Materials Science and Engineering A. 2009;**501**: 105-113. DOI: 10.1016/j.msea.2008.09.058

[42] Galiyev A, Kaibyshev R, Gottstein G. Correlation of plastic deformation and dynamic recrystallisation in magnesium alloy ZK60. Acta Materialia. 2001;**49**:1199-1207. DOI: 10.1016/S1359-6454(01)00020-9

[43] Figueiredo RB, Langdon TG. Strategies for achieving high strain rate superplasticity in magnesium alloys processed by equal-channel angular pressing. Scripta Materialia. 2009; **61**:84-87. DOI: 10.1016/j.scriptamat.2009.03.012

[44] Mukherjee AK. The rate controlling mechanisms in superplasticity. Materials Science and Engineering. 1971;**8**:83-89. DOI: 10.1016/0025-5416(71)90085-1

[45] Száraz Z, Trojanová Z, Zehetbauer M. High-pressure torsion deformation of a magnesium-based nanocomposite. International Journal of Materials Research. 2009;**100**:906-909. DOI: 10.3139/146.110126

[46] Langdon TG. An evaluation of the strain contributed by grain boundary sliding in superplasticity. Materials Science and Engineering A. 1994;**174**:225-230. DOI: 10.1016/0921-5093(94)91092-8

[47] Lee SW, Chen YL, Wang HY, Yang CF, Yeh JW. On mechanical properties and superplasticity of Mg-15Al-1Zn alloys processed by reciprocating extrusion. Materials Science and Engineering A. 2007;**464**:76-84. DOI: 10.1016/j.msea.2007.01.119

[48] Frost HJ, Ashby MF. Deformation-Mechanism Maps: The Plasticity and Creep of Metals and Ceramics. 1st ed. Oxford, UK: Pergamon Press; 1982. p. 166. ISBN: 978-0-08-029338-7

[49] Arzt F, Ashby MF, Verral RA. Interface controlled diffusion creep. Acta Metallurgica. 1983;**31**:1977-1989. DOI: 10.1016/0001-6160(83)90015-9

[50] Ryspaev T, Trojanová Z, Padalka O, Wesling V. Microstructure of superplastic QE22 and EZ33 magnesium alloys. Materials Letters. 2008;**62**:4041-4043. DOI: 10.1016/j.matlet.2008.05.063

Chapter 6

Discharge Behavior of Water-Activated Magnesium Battery

Isao Nakatsugawa, Yasumasa Chino and
Hideki Nakano

Additional information is available at the end of the chapter

http://dx.doi.org/10.5772/intechopen.79789

Abstract

Water-activated magnesium batteries possess several favorable attributes for energy storage and lighting sources. In this chapter, a portable-sized magnesium battery which consisted of MnO_2 cathodes and magnesium alloy anodes, being activated by drops of water, was investigated. The anode and cathode potential as well as their electrochemical impedance under a constant current load was monitored to evaluate the effect of water content, cell stacking, and discharging cycles. It was revealed that the discharge behavior was initially controlled by the depletion of water, followed by the cease of cathode reaction and the accumulation of $Mg(OH)_2$ at anode. The problem of low anode efficiency caused by stacking cells was analyzed, and its countermeasure was proposed. Several approaches to improve the battery performance were also presented.

Keywords: magnesium, primary battery, water-activated, electrochemical impedance, MnO_2

1. Introduction

Magnesium has a negative standard potential of −2.37 V vs. SHE and a high Faradaic capacity of 2.205 Ah/g, which is an attractive feature as the primary battery anode [1]. Mg primary batteries (dry batteries) were successfully used for military applications till 1970s [2]. Water/seawater-activated Mg batteries, or reserved batteries, are preferentially used for undersea devices or emergency signals [3]. Research and development of Mg-ion secondary batteries are under way. Charge-discharge performance comparable to lithium-ion batteries is a challenging task [4, 5]. In recent years, water-activated Mg batteries have been reevaluated [6–8].

Unlike other primary batteries, Mg batteries can be activated with neutral electrolyte when in use. This ensures a long shelf life, easy-to-use operation, and safe disposal of used batteries. Conventional sea water-activated Mg batteries use metal chloride cathodes [1, 3]. Recent water-activated Mg batteries follow the same principle, with incorporating modern electric and battery technologies. There are roughly two kinds of water-activated batteries. Type I battery consists of Mg anode, air-diffusion cathode, and electrolyte container. An appropriate amount of neutral electrolyte, usually NaCl solution, is injected in use. It is also called Mg-air battery [1, 6, 7]. Type II battery is a stacked structure of unit cells, and a small amount of water (1~2 ml) is injected in use [8]. Type I batteries can generate current densities of around tens of mA/cm^2 and can be served as small-middle sized generators. Although the capacity of the type II battery is limited (about 1 mA/cm^2), it is compact and boosts up easily by stacking unit cells, suitable for power saving devices like LED. There are several articles about type I battery [1, 6, 7], but technical information is limited for type II battery. Hereafter, type II battery is called as water-activated magnesium battery (WAMB) and described in this article.

Figure 1 shows several WAMB-operated LED lamps and their structure [9]. It is the size of a cigarette lighter, weighs about 20 g. A WAMB contains 3–4 unit cells stacked in a plastic casing. One or two LEDs are connected to the battery terminals. Details of the unit cell will be discussed later. Small amount of water is injected to the bottom of the casing to illuminate LED [10]. In the absence of water, juice and saliva can be used. As water is gradually lost during discharge, its intermittent supply is necessary. **Figure 2** shows the change in LED illuminance with water injection [9]. Comparisons with other illumination sources are mentioned. Each water injection restores illuminance while the intensity gradually disappears. After 6–7 injections, the WAMB cannot generate enough power for the LED and goes out. One WAMB can provide light in about a week, which is sufficient as an emergency illumination source. In addition to its compactness, it does not contain any hazardous substances, which enables safe operation and disposal of the used WAMB [8, 9].

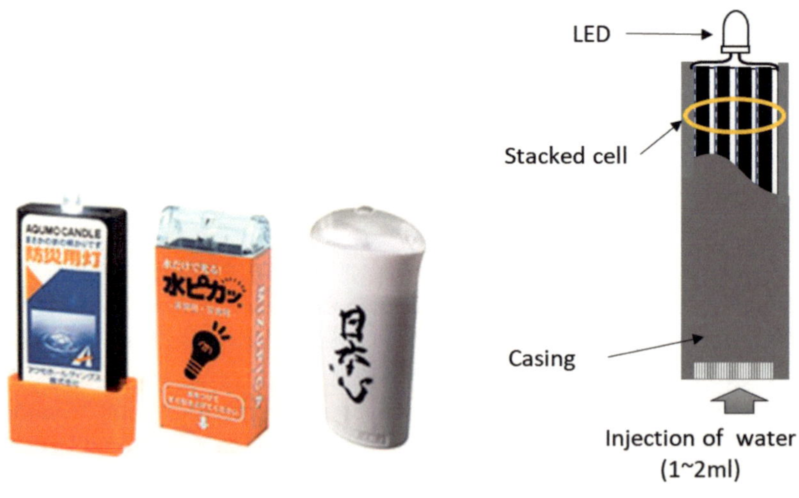

(a) (b)

Figure 1. Appearance and structure of type II WAMB. (a) Commercial product (handy LED lamp). (b) Structure.

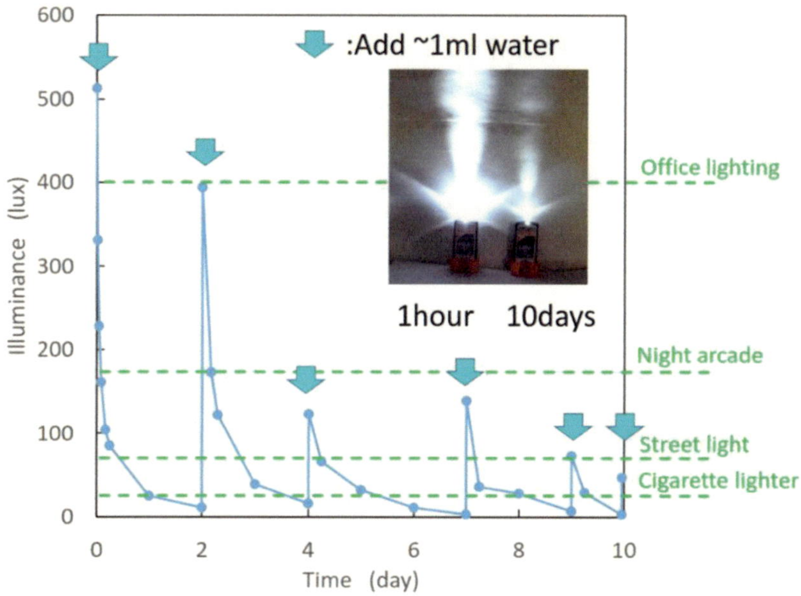

Figure 2. Illuminance of WAMB (stacked cell). Illuminance was measured at 30 cm distance from LED [9].

In this chapter, the discharge performance of WAMB is presented based on our recent research [11–13]. We first describe the discharge behavior of unit cells, following the studies of cell stacking. Factors that affect discharge performance are identified and analyzed. Several approaches to improve performance are also proposed.

2. Materials and methods

A Mg-3%Al-1%Zn alloy (AZ31) magnesium alloy plate having a thickness of 0.8 mm (manufactured by Osaka Fuji Corp.) was used as the anode. The chemical composition is shown in **Table 1**. The surface was polished with #600 SiC paper, rinsed with water, and wiped with ethanol. As a cathode material, a part of commercially available WAMB (Aqumo Holdings "Aqumo candle®") [9] was used. Although details are not clarified, it seems to consist of pulp-based nonwoven fabric containing MnO_2, catalyst, graphite, and NaCl [8]. A Cu foil having a thickness of 0.6 mm was used as the current collector of the cathode. As the separator, pulp/polypropylene nonwoven fabric having a thickness of 0.25 mm was used. These electrodes were piled in the order of current collector/cathode/separator/anode to construct the unit cell, shown in **Figure 3**. The electrode size was 25 × 50 mm except for the separator which had 10 mm longer than others for absorbing water. A Pt wire of ϕ 0.6 mm was sandwiched

	Al	Zn	Mn	Si	Fe + Cu + Ni	Mg
AZ31B	3.03	0.78	0.29	0.021	<0.006	Bal.

Table 1. Chemical composition of the Mg anode.

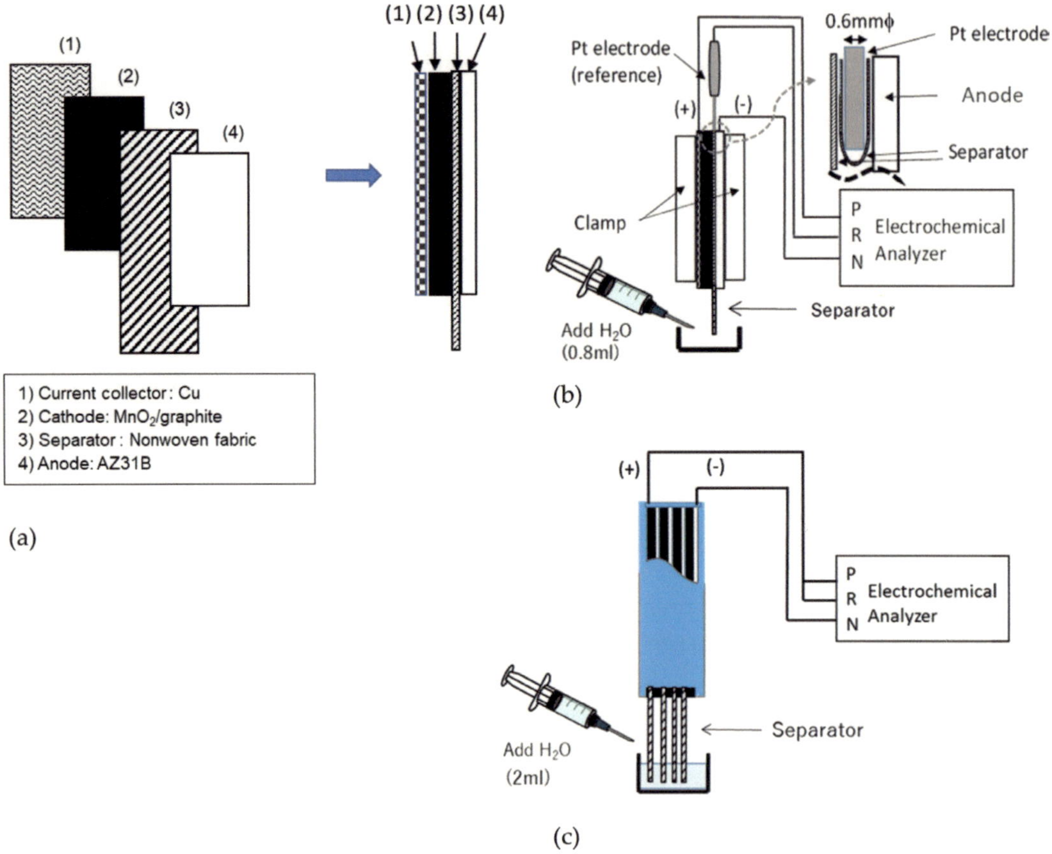

Figure 3. Experimental setup of WAMB discharge study. (a) Structure of the unit cell. (b) Setup of the unit cell. (c) Setup of the stacked cell.

between the separator and the anode and used as a reference electrode. To avoid direct contact with the Pt wire, a small piece of the separator was attached to the anode side. Two Cu sheets were attached to the outside of the unit cell and used as terminals. The assembly was further cramped with two acrylic plates under a pressure of 6 N. Stacked cells were prepared by superimposing four unit cells in series and inserted into a commercial WAMB casing. Tap water (Sagami river system, Samukawa water purifying plant) was used to activate WAMB by injecting them into the bottom of the separator. The amount of water was varied from 0.4 to 0.8 ml in a unit cell and 2 ml in a stacked cell.

The initial discharge capacity of WAMB was analyzed with a unit cell by measuring a potentiodynamic polarization curve. After 20 min of water injection, the cell voltage V_{cell} and the discharge current density i_d were monitored at a scanning rate of 1 mV/s.

A constant current discharge test of 0.24 mA/cm^2 in the unit cell and 0.08–0.80 mA/cm^2 in the stacked cell was performed. The anode potential E_a, the cathode potential E_c, V_{cell}, and electrochemical impedance spectroscopy (EIS) of the anode, cathode, and cell were independently measured in the unit cell. In the stacked cell, V_{cell} and the cell impedance were monitored. The EIS was performed in a current-controlled mode with an amplitude of 1% of i_d, and the scanned frequency was changed from 100 kHz to 1 Hz using an electrochemical apparatus (Solartron Analytical, Modulab Xm). The obtained spectrum was analyzed by curve fitting software (Scribner ZView). When V_{cell} fell below the terminal voltage of 0.7 V in the unit cell and 2.5 V in the stacked cell, the discharge test was interrupted. After that, the same amount of water was injected, and the test was restarted. When V_{cell} decreased to the terminal voltage within 1 h after water injection, the discharge test was terminated, and cumulative discharge time t_{total} (h) was calculated.

The discharge capacity was evaluated by the gravimetric energy density E_g (Wh/kg).

$$E_g = V_{cell.avg} \times i_d \times S \times t_{total} / (M + L) \quad (1)$$

$V_{cell.avg}$ is the average of the monitored V_{cell}, S is the area of the electrode (= 12.5 cm^2), M is the cell weight (g), and L is the total amount of injecting water (g). M does not include the weight of the casing. For the calculation of Eq. (1), L is fixed as 1.5 g for a unit cell and 6 g for a stacked cell, referring the result of the weight increase after the discharge test.

After the test, the discharged cells were disassembled, and the appearance of the electrodes was examined. The consumption of Mg anode ΔW (g) was measured after removing the discharged product by a wire brush. Current efficiency CE (%) was calculated by Eq. (2).

$$CE = i_d \times S / I_{grav.} \times 100 \quad (2)$$

Igrav. (mA) was calculated with ΔW, assuming Mg is dissolved with the ionic valency of 2.

$$I_{grav.} = 2 \times 96{,}500 \times \Delta W / t_{total} / 24.31 = 2206\ \Delta W / t_{total} \quad (3)$$

3. Results

3.1. Initial discharge capacity of unit cell

Figure 4 shows the polarization curve of the unit cell with a water injection of 0.8 ml [13]. The power density reached about 10 mW/cm^2 with i_d of 15 mA/cm^2. As described later, the current range used in the constant discharge test was less than 1 mA/cm^2 (<10 mA), and V_{cell} was about 1.4 to 1.5 V. By stacking four cells in series, V_{cell} reaches 5.6 ~ 6 V, which is sufficient to illuminate a small LED with a forward voltage of 2.5–3.5 V.

3.2. Discharge behavior of unit cell with water injection

Figure 5(a) shows the discharge curve of WAMB unit cell at the initial injection of water [11]. The amount of water was changed from 0.4 to 0.8 ml. The unit cell did not hold more than 1 ml of water. V_{cell} was kept at about 1.4 V for a certain period and declined rapidly. The larger the amount of water, the longer the discharge time was obtained. **Figure 5(b)** shows the dependence of the discharge time on the amount of water. A relationship expressed by a quadratic function can be seen. Hereafter, the injection of water was fixed to 0.8 ml.

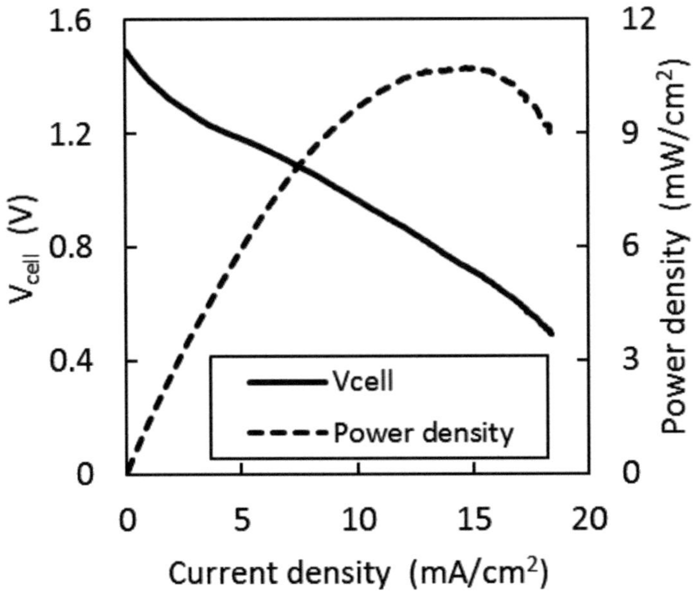

Figure 4. Initial discharge performance of WAMB (unit cell) [13].

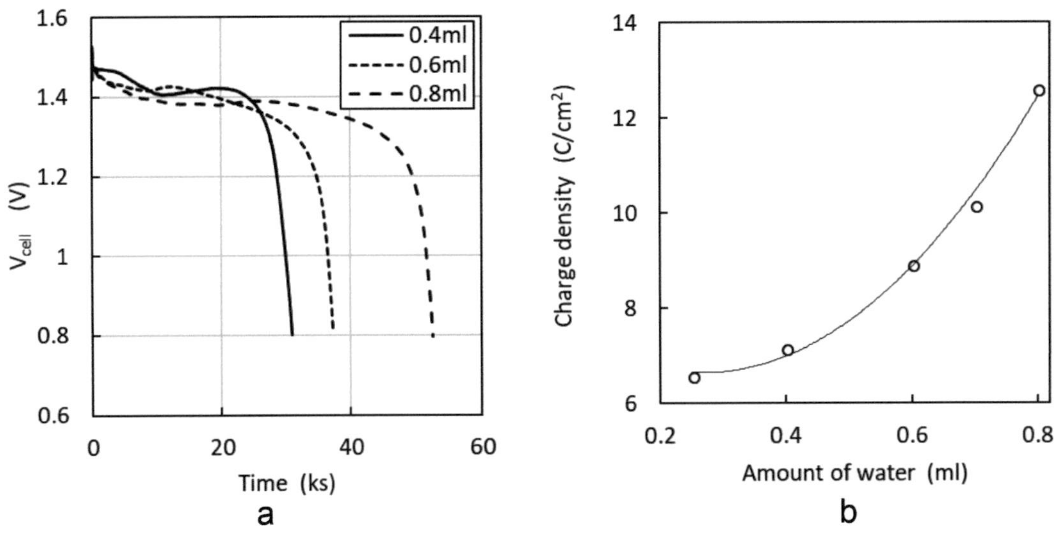

Figure 5. Discharge behavior of WAMB (unit cell) at the first injection of water at a current density of 0.24 mA/cm² [11]. (a) Discharge curves of WAMB. (b) Dependence of discharged period on the water content.

Figure 6 presents the discharge behavior of WAMB with different discharge cycles [13]. E_a kept constant about −1.6 V regardless of discharge time and cycles, and E_c changed every discharge cycle. V_{cell} followed the change in E_c. The discharge time was gradually shortened by injection cycles shown in **Figure 6(b)**. It was possible to regenerate WAMB up to 8 times of water injection. t_{total} was calculated to be 51.7 h.

Figure 7 exhibits the change of Nyquist diagram of the anode, cathode, and unit cell with different injection cycles. The Mg anode drew a depressed capacitive semicircle, followed by an inductive loop in the low frequency region. In general, the size of trajectory increased with injection cycles. The impedance of the cathode was negligible in the first cycle but increased thereafter. Furthermore, a Warburg impedance was recognized in the low frequency region.

There are several EIS studies on the corrosion of Mg or Mg alloys [14–16], but the application to neither Mg battery nor Mg anodized in NaCl solution is rare. Then, referring the equivalent electric circuit of pure Mg corroding in NaCl solution [14], a circuit for the WAMB anode is proposed in **Figure 8(a)** [13].

$R_{s.a}$: Solution resistance at the anode.

R_{1a}, R_{2a}: Resistance associated with the dissolution of the anode.

R_{3a}: Resistance associated with the hydrogen gas evolution on the anode.

CPE_{1a}: Constant phase element associated with the discharge product on the anode.

CPE_{2a}: Constant phase element associated with the electrochemical double layer of the anode.

L_a: Inductance associated with the dissolution of the anode.

As for the WAMB cathode, an equivalent circuit was expressed in **Figure 8(b)**, referring the MnO_2 electrode of Zn-MnO_2 battery [17, 18].

$R_{s.c}$: Solution resistance at the cathode.

$R_{ct.c}$: Resistance associated with the reduction of MnO_2.

CPE_c: Constant phase element associated with the electrochemical double layer of the cathode.

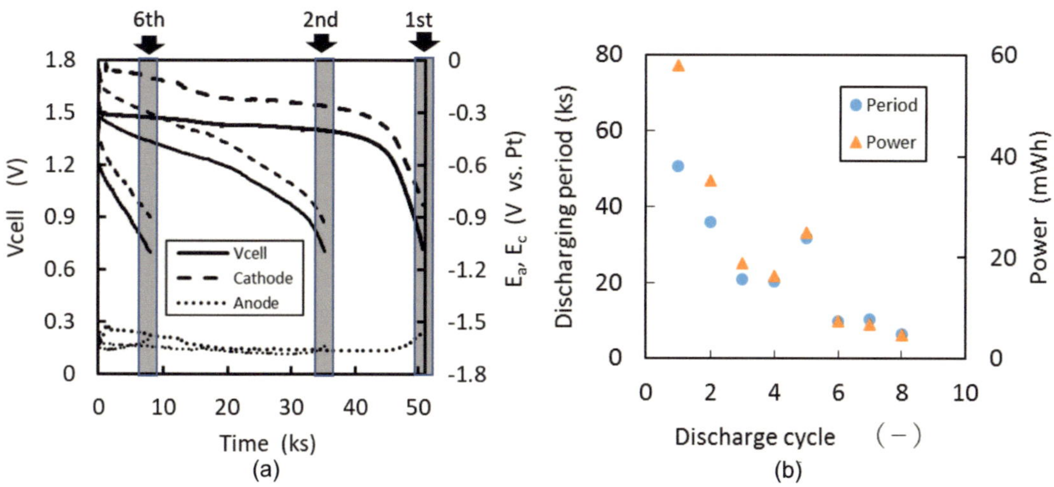

Figure 6. Discharge behavior of WAMB (unit cell) at a current density of 0.24 mA/cm² [13]. (a) Change of E_a, E_c, and V_{cell} with time under different discharge cycle. (b) Change of discharge time and power with the discharge cycle.

W_c: Warburg impedance of the cathode (open).

Here, constant phase element (CPE) [19] such as CPE_{1a}, CPE_{2a}, and CPE_c was employed instead of capacitance to reflect the inhomogeneity of the electrode surface.

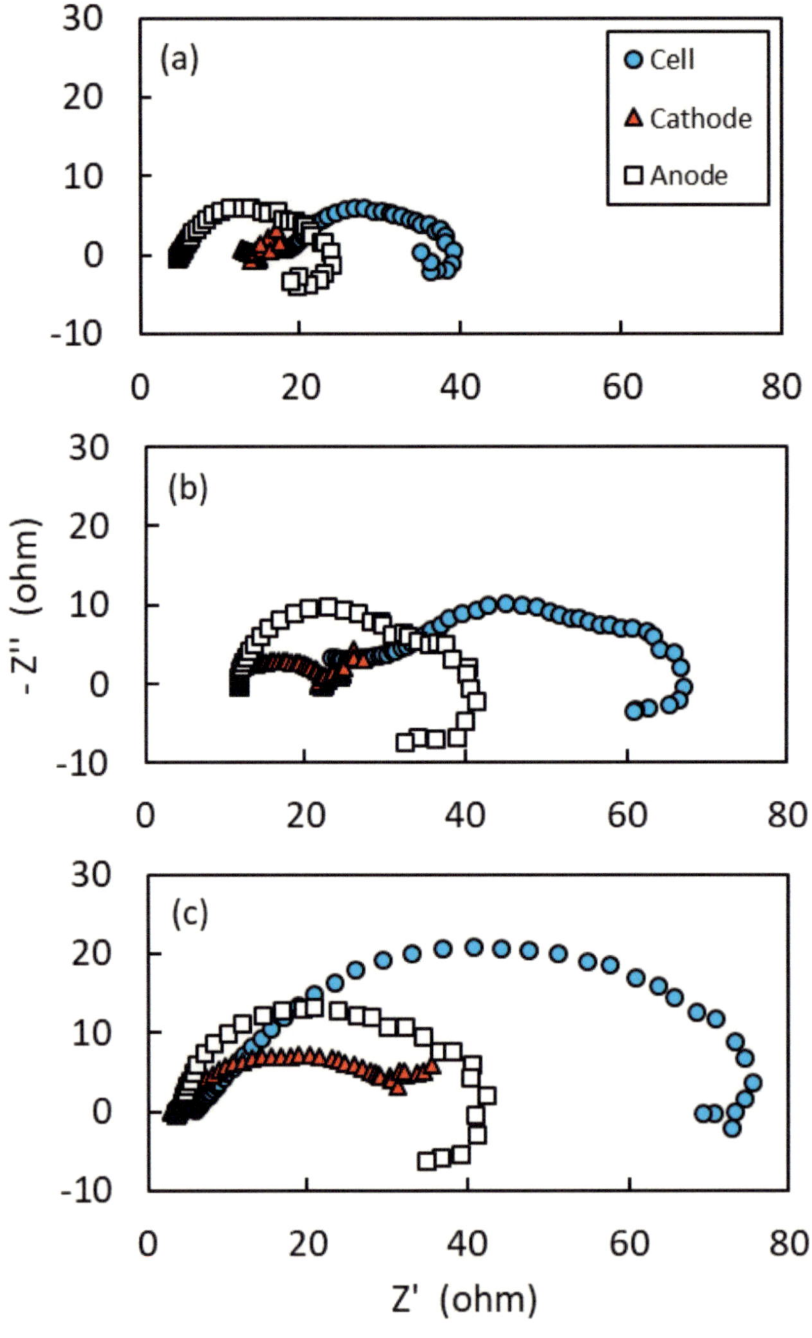

Figure 7. Nyquist diagrams of the anode, cathode, and cell of WAMB unit cell after (a) first, (b) second, and (c) sixth cycles of water injection [13].

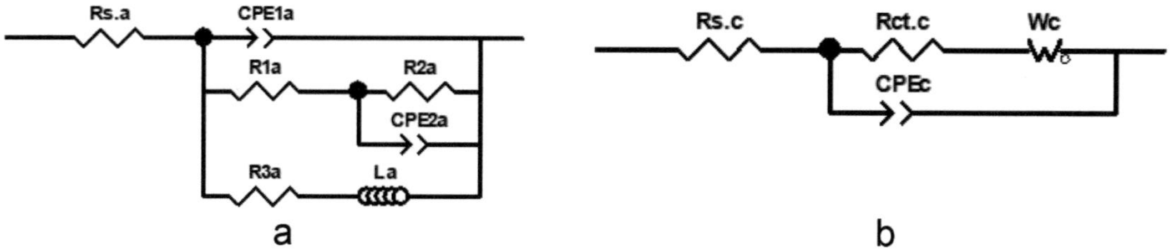

Figure 8. Electrical equivalent circuits for the WAMB anode and the WAMB cathode. (a) Anode. (b) Cathode.

Anode									
	$R_{s.a}$ (Ω)	R_{1a} (Ω)	R_{2a}(Ω)	R_{3a} (Ω)	$CPE_{1a\text{-}T}$ (-)	$CPE_{1a\text{-}}$(-)	$CPE_{2a\text{-}T}$ (-)	$CPE_{2a\text{-}P}$ (-)	L_a (Ωs)
First	4.47	3E-07	23.3	27.2	3.74E-03	0.284	2.01E-04	0.84	4.17
Second	11.3	0.13	36.1	34.8	1.93E-03	0.424	9.13E-06	1.14	6.96
Sixth	3.05	3.72	36.5	58.7	4.01E-04	0.733	2.85E-06	1.25	12.9
Cathode									
	$R_{s.c}$ (Ω)	$R_{ct.c}$ (Ω)	$CPE_{c\text{-}T}$ (-)	$CPE_{c\text{-}P}$ (-)	$W_{c\text{-}R}$ (-)	$W_{c\text{-}T}$ (-)	$W_{c\text{-}P}$ (-)		
First	12.3	1.71	8.99E-06	0.956	9.7	0.702	0.608		
Second	11.6	10.3	2.90E-05	0.744	17.6	1.053	0.744		
Sixth	2.24	25.7	4.49E-04	0.628	10.8	0.393	0.628		

Table 2. Parameters of equivalent circuit shown in **Figure 8** to simulate the response of WAMB discharged at 0.24 mA/cm^2.

The curve fitting of the experimental data of each electrode was performed using simulation software. The results are shown in **Table 2** and superimposed on the experimental data in **Figure 9**. The proposed circuit approximated the impedance spectrum, showing the depressed capacitive semicircle, the presence of the inductive loop at the anode, and the Warburg impedance at the cathode. Curioni et al. [15] reported that the reciprocal of (R_{1a} + R_{2a}) shows a good correlation with the corrosion rate of pure Mg obtained from the hydrogen evolution reaction. Then, the parameter (R_{1a} + R_{2a}) is replaced with $R_{ct.a}$, the sum of $R_{ct.a}$ and $R_{ct.c}$. is replaced with $R_{ct.cell}$, and discussed herewith. These parameters and R_s at each electrode can be obtained roughly from the trajectory shown in **Figure 9** [14].

Figure 10 plots the time dependence of R_s and R_{ct} of the anode, cathode, and cell at different injection cycles. In general, R_s increases with the discharge time, suggesting a decrease in the water content in the cell. However, at the sixth water injection, V_{cell} reached the terminal voltage before R_s started to increase. In case of R_{ct}, the cathode took a smaller value at the initial stage than the anode. After that, both values increased with the number of water injections. In the sixth water injection, $R_{ct.a}$ and $R_{ct.c}$ reached the same level.

Figure 11 presents the potentiodynamic polarization curves of the cathode before and after the discharge test. The cathode current decreased significantly after the test, suggesting a decrease in discharge capacity. In case of the anode, a white-gray discharge product was observed. After removing them, the thickness of the Mg plate was partly diminished, but the shape remained.

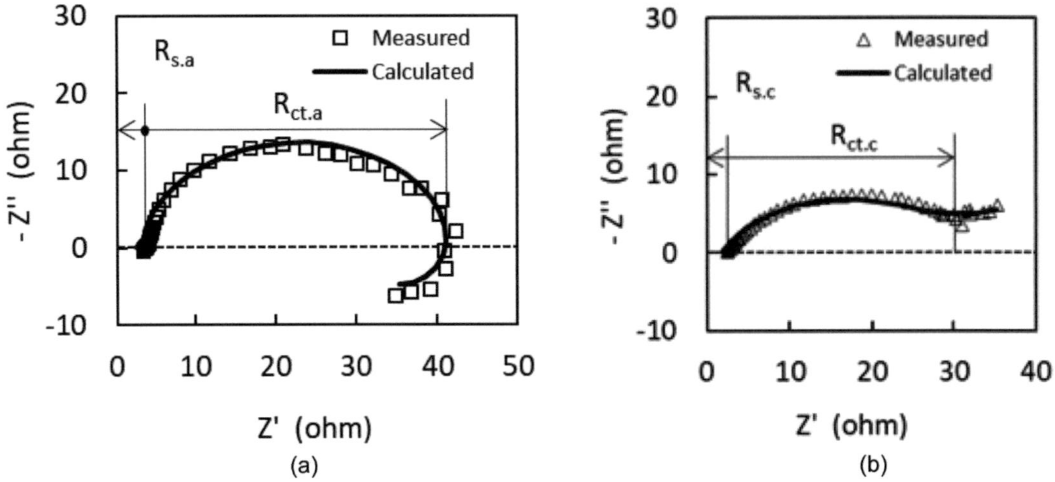

Figure 9. Comparison between the calculated and the measured impedance spectra of WAMB. (a) Anode after 6th cycle. (b) Cathode after 6th cycle.

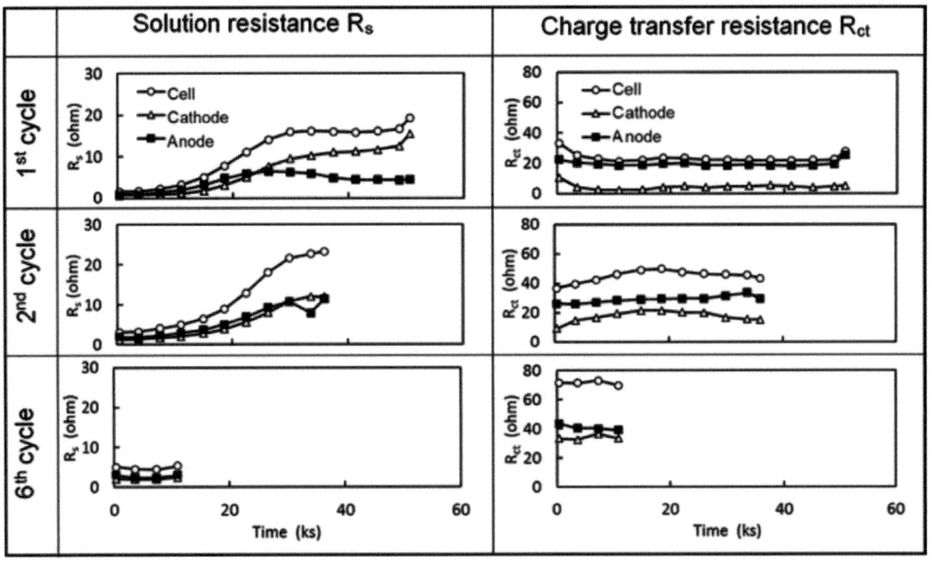

Figure 10. Time dependence of R_s and R_{ct} of the WAMB cathode and the WAMB anode and their combined cell (unit cell) at a current density of 0.24 mA/cm^2.

3.3. Discharge behavior of stacked cell with water injection

Time variation of V_{cell} of the stacked WAMB at a current density of 0.24 mA/cm^2 was presented in **Figure 12**. As expected, V_{cell} has quadrupled by stacking four unit cells. The behavior of V_{cell} was basically the same as that of the unit cell. Total discharge time reached 95.6 h, which was about twice that of the unit cell. **Figure 13** shows the Nyquist diagram of the stacked

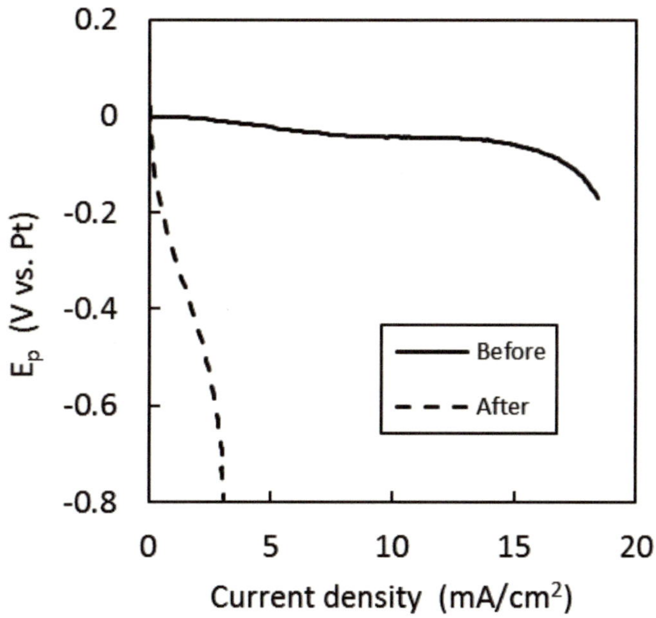

Figure 11. Potentiodynamic polarization curves of the WAMB cathode (unit cell) before discharging and after six cycles of discharging.

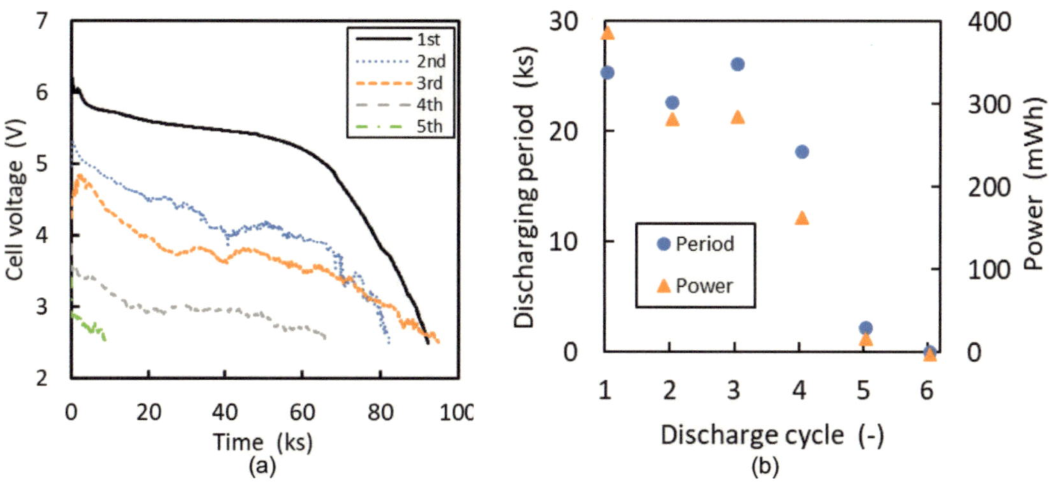

Figure 12. Discharge behavior of the WAMB stacked cell at a current density of 0.24 mA/cm². (a) Discharge curves. (b) Change of discharge time and power with discharge cycle.

cell after the first, third, and fifth water injections. The spectra drew depressed capacitive semicircles up to the fourth water injection. The diameter (= $R_{ct.cell}$) was initially large, but gradually became smaller. In the fifth water injection, a Warburg impedance appeared in the low frequency region. $R_{s.\,cell}$ and $R_{ct.cell}$ at the terminal voltage were 24 and 288 Ω, respectively, which were about four times the value in the unit cell.

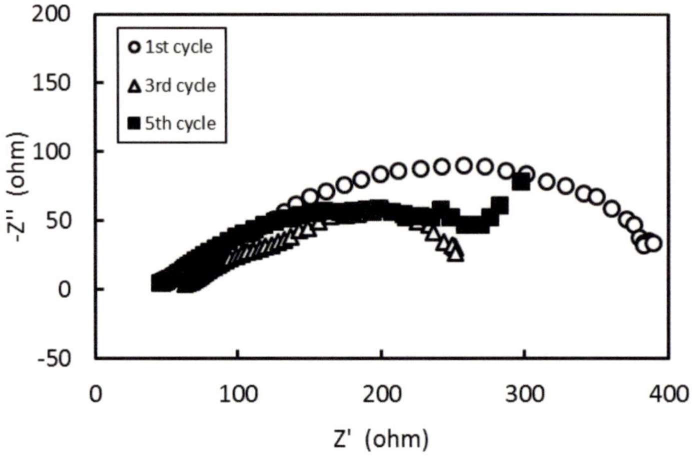

Figure 13. Impedance spectra of WAMB (stacked cell) at the termination voltage after different discharging cycles.

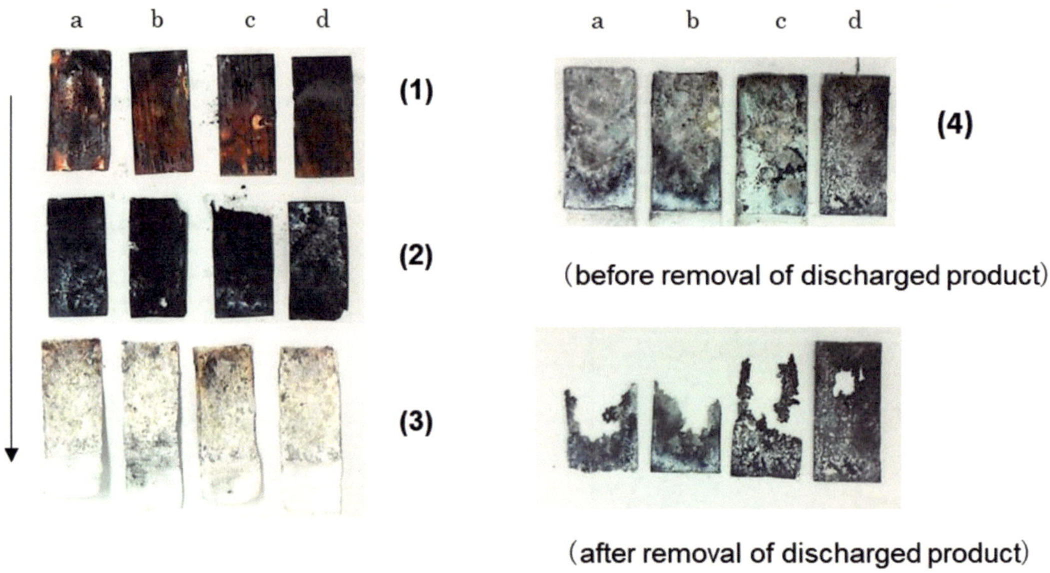

Figure 14. Appearance of WAMB (stacked cell) at a current density of 0.24 mA/cm^2 after six cycles of discharge [13].

3.4. Discharge capacity and current efficiency

The appearance of the electrode after the discharge test was shown in **Figure 14**. They are: (1) current collector, (2) cathode, (3) separator, and (4) anode before/after the removal of the discharge product. The alphabetical suffix indicates the layer of the stacked cell. Thus, the first unit cell was stacked in the order of *[a1/a2/a3/b4]*, following the second cell as *[b1/b2/b3/b4]*, the third as *[c1/c2/c3/c4]*, and the fourth as *[d1/d2/d3/d4]*. At the anode, a white and gray discharge product was visible. After the removal of the product, more than half of the anode

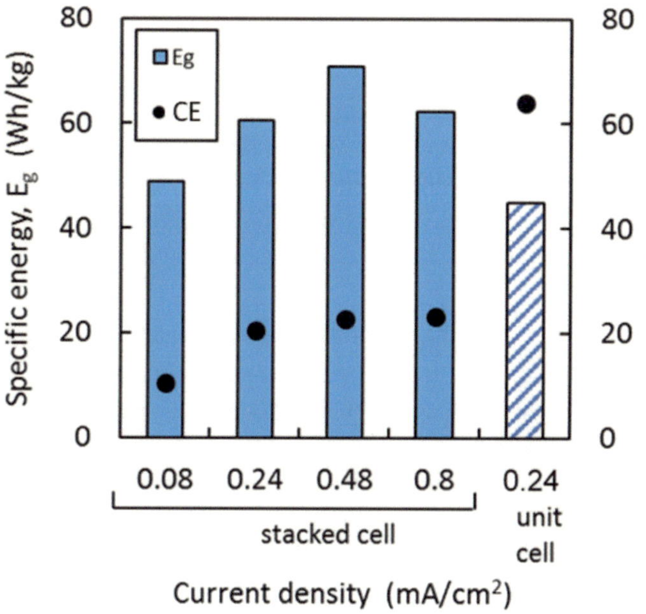

Figure 15. Specific energy and current efficiency of WAMB [13].

was disappeared except for the *d4*, which was located on the outermost side of the stacked cell. The surface of the Cu current collector attached to the Mg anodes was blackened.

Figure 15 illustrates the energy density E_g and the current efficiency *CE* of the unit, and the stacked cell. E_g reached its maximum of 71 Wh/kg at i_d of 0.48 mA/cm^2. It is noticeable that *CE* was greatly decreased by stacking the unit cells.

4. Discussions

4.1. Discharge behavior of WAMB unit cell

Discharging of WAMB begins with feeding water to the cathodes via separators. Water dissolves NaCl in the cathode material and creates an ionic path between the electrodes. The cell reactions are expressed as:

$$\text{Anode:} Mg \rightarrow Mg^{2+} + 2e \quad (4)$$

$$\text{Cathode:} 2MnO_2 + H_2O + 2e \rightarrow Mn_2O_3 + 2OH^- \quad (5)$$

$$\text{Total:} Mg + 2MnO_2 + H_2O \rightarrow Mn_2O_3 + Mg(OH)_2 \quad (6)$$

Eq. (6) is an exothermic reaction. Enthalpy change $-\Delta H$ was calculated to be 484 kJ/mol [20]. Since WAMB is an open structure, the injected water evaporates gradually due to heat. Water was also consumed by Eq. (5) to form $Mg(OH)_2$. Therefore, intermittent water injection is indispensable. The discharge behavior of WAMB depends on the dissolution of the Mg anode and the reduction of MnO_2 in the cathode. As shown in **Figure 6**, E_a was stable regardless of the discharge time and the cycle of water injection. In contrast, E_c varied with discharge conditions that affected V_{cell}. All these facts suggest that WAMB is the cathode-controlled system. The decline in the cathode capacity after the discharge shown in **Figure 11** confirms such an assumption.

Based on the EIS result shown in **Figure 10**, the anode and cathode behaviors during the discharge can be schematically interpreted as **Figure 16**. At the initial stage, $R_{ct.c}$ is smaller than $R_{ct.a}$ and does not affect the discharge behavior. As time elapses, $R_{s.cell}$ gradually increases due to the depletion of water and stops the discharge reaction. After repeated water injection, the initial $R_{ct.c}$ and $R_{ct.a}$ tend to show larger values than the preceding cycle. This suggests a decrease in the conductive MnO_2 in the cathode and a decrease in the reaction area of the anode due to the accumulation of $Mg(OH)_2$. In addition, a decrease in E_c lowers V_{cell}. At the end of the period, even if the water amount is sufficient (small $R_{s.cell}$), the gap between V_{cell} and the terminal voltage becomes small, so the discharge ends in a short time.

At the terminal stage, the Warburg impedance appeared in the cathode of **Figure 7(b)** and **(c)**. Qu [18] discussed the existence of Warburg impedance appearing at the MnO_2 cathode of an alkaline Mn battery system, in terms of the volume expansion of the cathode during discharge. Such expansion prevents the proton diffusion of the entire lattice of MnO_2, which is an important rate-determining step. When the discharge exceeds 50%, the localized Jahn-Teller deformation begins to occur and the proton diffusion coefficient decreases. Though such volume expansion was not recognizable in the present system, structural analysis of the cathode using XRD would provide detailed mechanism of the degradation of WAMB.

4.2. Discharge behavior of WAMB stacked cell

In general, the discharge behavior of the stacked cell follows that of the unit cell. It is worth noting that t_{total} almost doubled by stacking. The exact cause is not clear, but it may be related to the generation of the voluminous $Mg(OH)_2$ which is 73.5% more molar volume than Mg metal [21]. Such expansion compresses the electrodes in the casing, improves mutual contact, and would lead to the prolongation of the discharge time.

On the other hand, *CE* was sharply decreased in the stacked cell. *CE* of 64% obtained at the unit cell is within the range of 55–66%, which is reported in the corrosion of Mg [22]. The reason that the *CE* of Mg is less than 100% in Mg is closely related to the hydrogen gas evolution reaction that occurs simultaneously even if Mg is anodically polarized [23, 24].

$$2H_2O + 2e \rightarrow H_2 + 2OH^- \quad (7)$$

However, the fact is that *CE* as low as 20% in the stacked cell cannot be explained in this context. The reason for this is interpreted by galvanic corrosion by the Cu current collector, which

http://dx.doi.org/10.5772/intechopen.79789

is illustrated in **Figure 17** [13]. During water injection, most of the water penetrates from the separator into the cathode material osmotically. However, some of them may penetrate small gaps between the Cu collector and the Mg anode. As a result, a galvanic cell having Mg as anode and Cu as cathode is formed, and the former is corroded preferentially. In this case, the anode reaction is represented by Eq. (4) and the cathode reaction is by Eq. (7). In principle, WAMB establishes the flows of electron and ions (OH^-) and generates electricity as shown in Section (a) of **Figure 17**. However, the portion where the galvanic corrosion is being occurred forms the closed circuit shown in section (b), and the Mg anode does not contribute to the discharge performance and is wasted by corrosion. The fact that the Mg anode *d4* in **Figure 14** is less susceptible to corrosion than other anodes corroborates the absence of such galvanic corrosion. In addition, galvanic corrosion results in forming the corrosion products between the anode and the Cu collector, causing IR drop and reducing V_{cell}. Cu collectors are also oxidized (see **Figure 14**—(1)) to increase the contact resistance. To reduce the loss due to galvanic corrosion and maximize the discharge capacity of WAMB, the choice of higher i_d up to around 0.5 mA/cm^2 is recommended.

4.3. Technical challenge and future applications of WAMB

Based on current research, the gravimetric energy density E_g of WAMB is 71 kWh/kg, which is inferior to Zn-alkaline MnO_2 battery (154 Wh/kg) or Zn-air battery (415 Wh/kg) [25]. Since WAMB inherits the dry Mg-MnO_2 battery or the reserved battery technologies [2, 3], of which E_g is less than 100 Wh/kg [25], dramatic improvement of WAMB may not be feasible. Still, there are several rooms to enhance the capabilities of WAMB.

A. Design to eliminate galvanic corrosion in the stacked cell.

B. Catalyst for enhancing the ability of the cathode [26–30].

C. Mg alloy for improving discharge capacity [31–33].

D. Selection of electrolyte and additive [34–37].

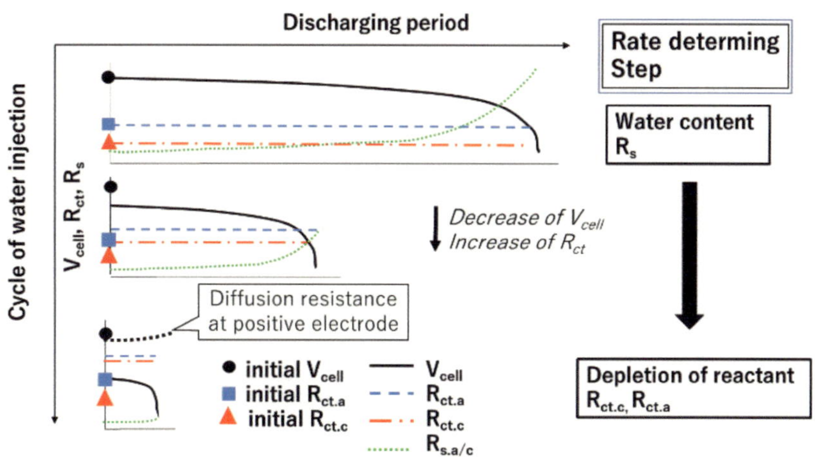

Figure 16. Schematic diagram of the time variation of electrochemical parameters in WAMB [13].

They are briefly commented as follows. Several studies conducted in type I water-activated Mg battery are also useful for type II WAMB. In addition, we performed some experimental trials that are summarized in **Table 3**. The relative increase in t_{total} compared to the standard condition was used as an index of improvement.

4.3.1. Design for eliminating galvanic corrosion in the stacked cell

The galvanic corrosion occurs in the condition of stacking unit cells, and it seems that the penetration water in the gap between the anode and the Cu collector is the main cause. It is conceivable to separate the unit cells and connect them outside the casing to avoid the galvanic contact. Although it is technically possible, the structure becomes complicated. A solution is to bond the electrodes using conductive tape to seal the gap. Although it induces contact resistance to decrease V_{cell}, it is effective to eliminate the galvanic corrosion and improve the *CE*. With this configuration, a 150% increase in t_{total} was possible in our preliminary trials.

4.3.2. Catalyst for enhancing the ability of the cathode

Eq. (5) can be considered as a kind of oxygen reduction reaction. Therefore, the addition of oxygen catalyst shifts E_c more noble. Apart from Pt group catalysts, several organic [26, 27], inorganic [28], or complex [29, 30] catalysts are promising. We studied the effect of Ni, Cu, and Fe powder in the MnO_2 containing cathode mix. The addition of Ni powder gave the best performance among them.

4.3.3. Mg alloy for improving discharge capacity

In this study, a versatile AZ31 alloy was used as the anode. There are several Mg alloys developed for type I water-activated Mg batteries. It is effective to improve the discharge capacity

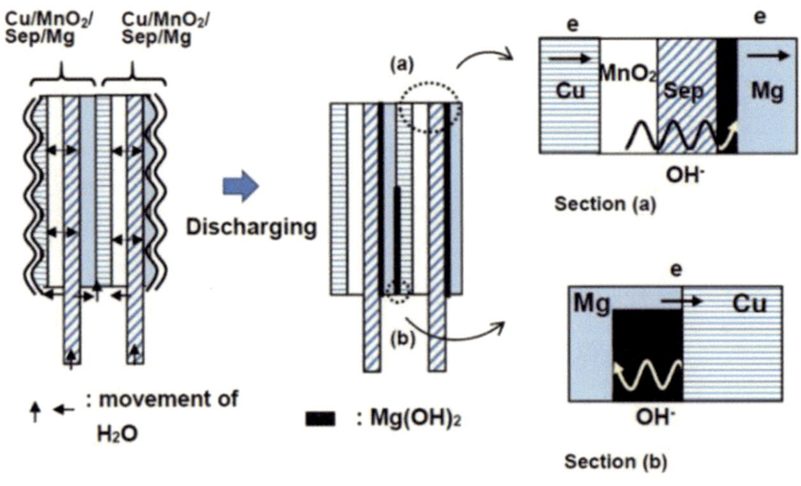

Figure 17. Schematic diagram of electron and ionic flows of WAMB (stacked cell): Section (a) indicates the normal discharge behavior, and section (b) indicates the abnormal behavior caused by the occurrence of a galvanic corrosion between the Cu current collector and the Mg anode.

Method	Modification	Cell	t_{total} (%)
A	Cu plated conductive tape bet. Mg/Cu	Stack	146
B	Add Cu powder in cathode mix	Stack	82
	Add Ni powder in cathode mix	Stack	135
	Add Fe powder in cathode mix	Stack	105
C	Use Mg-Li-Al alloy as anode	Unit	208
D	Add LiCl in cathode mix	Stack	163
—	None	—	100

Table 3. Technical proposals to improve the discharge performance of WAMB.

by adding specific alloying elements such as Li [31], Ca [32], Pb [33] to Mg-Al alloys. As shown in **Table 3**, it was found that Mg-14Li-1Al alloy demonstrated excellent performance as the anode. The presence of lithium hydroxide in the discharge product seems to maintain $R_{s.a}$ low even after multiple water injections. Similar effects are expected for Mg-Al alloys containing Ca [32].

4.3.4. Selection of electrolyte and additive

NaCl is the standard electrolyte for type I and II water-activated batteries. It results in a larger V_{cell} compared to a phosphate- or nitrate-based electrolyte but lowers *CE* [34]. Magnesium forms a soluble compound with organic acids such as acetic acid and citric acid [35]. Addition of citric acid did not improve the discharge performance of type I battery [36]. Some additives form chelates with Mg which raise V_{cell} [37]. We found that the use of LiCl instead of NaCl increased t_{total}. Although the reason is uncertain, there is a possibility that the same effect as the Mg-Li-Al alloy anode described above may have occurred.

Finally, we consider that the recent attention of WAMB comes from aspects other than power-oriented issues, that is, the convenience and safe handling, or expectations of Mg as an energy carrier for a sustainable society [38]. Apart from LED lamps, water sensors from diapers [39] to underground pipelines [40], biodegradable batteries for medical implants or home appliances [41, 42] can be expected as new applications of WAMB.

5. Conclusions

A water-activated magnesium battery (WAMB) consisting of MnO_2 cathode and magnesium anode was evaluated. The obtained results are as follows:

1. The initial power density of WAMB showed the maximum of 10 mW/cm^2 at 15 mA/cm^2. The cell voltage under a constant current discharge of 0.24 mA/cm^2 was 1.4–1.5 V over 50 h at the unit cell. Changes in the cell voltage were closely related to the performance of the cathode. The injection of water regenerated the WAMB unit cell up to 8 times.

2. Electrochemical impedance spectroscopy was applied to monitor the discharge performance of WAMB. An electrical equivalent circuit of the anode and the cathode was proposed from the frequency spectra. They consisted of solution resistance, charge transfer resistance, constant phase element, inductance for the anode, and Warburg impedance for the cathode. Curve fitting using simulation software agreed well the experimental data.

3. By monitoring solution resistance and charge transfer resistance, the degradation behavior of WAMB was analyzed. At the initial stage, an increase of solution resistance due to the depletion of water determined the discharge time. When water was repeatedly injected to regenerate WAMB, the increase in the charge transfer resistance of the anode and the cathode controlled the discharge time and the cell voltage, respectively. At the final stage, the Warburg impedance appeared in the cathode frequency spectrum. These data suggest that WAMB is the cathode-controlled system. The loss of discharge capacity and the volume expansion of MnO_2 was considered.

4. Stacking the unit cells in series was effective for increasing the cell voltage and extending the discharge time. On the other hand, the current efficiency decreased from 64 to 10–20%. The reason was interpreted in terms of a galvanic corrosion occurring at the Mg anode and the Cu current collector.

5. The power density of WAMB was 71 Wh/kg, which was lower than the current Zn batteries. Several approaches to improve performance were proposed. Promising applications of WAMB including biodegradable batteries were also mentioned.

Author details

Isao Nakatsugawa[1]*, Yasumasa Chino[1] and Hideki Nakano[2]

*Address all correspondence to: i.nakatsugawa@aist.go.jp

1 Structural Materials Research Institute, National Institute of Advanced Industrial Science and Technology, Nagoya, Japan

2 CLT Ltd., Nagoya, Japan

References

[1] Wang N, Wang R, Peng C, Hu C, Feng Y, Peng B. Research progress of magnesium anodes and their applications in chemical power sources. Transactions of the Nonferrous Metals Society of China. 2014;**242**:427-2439. DOI: 10.1016/S1003-6326(14)63367-7

[2] Spellman PJ. Magnesium and aluminum batteries. In: Reddy T, editor. Linden's Handbook of Batteries. 4th ed. New York: McGraw-Hill Education; 2011. pp. 10.1-10.13

[3] Lucero RD, Karpinski AP. Reserve Magnesium Anode and Zinc/Silver Oxide Batteries. In: Reddy T, editor. Linden's Handbook of Batteries. 4th ed. New York: McGraw-Hill Education; 2011. pp. 34.1-34.7

[4] Saha P, Datta MK, Velikokhatnyi OI, Manivannan A, Alman D, Kumta PR. Rechargeable magnesium battery: Current status and key challenges for the future. Progress in Materials Science. 2014;**66**:1-86. DOI: 10.1016/j.pmatsci.2014.04.001

[5] Yoo HD, Shterenberg I, Gofer Y, Gershinsky G, Pour N, Aurbach D. Mg rechargeable batteries: An on-going challenge. Energy & Environmental Science. 2013;**6**:2265-2279. DOI: 10.1039/C3EE40871J

[6] Zhang T, Tao Z, Jun C. Magnesium-air batteries: From principle to application. Materials Horizons. 2014;**1**:196-206. DOI: 10.1039/c3mh00059a

[7] Ito A, Abe H, Kubota M, Taira Y, Iizuka H, Saito K, Takahara T, Ono Y. Evaluation test of magnesium air battery 'Mg box' for emergencies. FB Technical News. 2014;**70**:23-27

[8] Suzuki S. JP Patent No. 519613, WO2013/018769

[9] Aqumo Candle [Internet]. 2017. Available from: http://www.aqumo-holdings.com [Accessed: May 24, 2018]

[10] Aqumo Candle [Internet]. 2015. Available from: https://www.youtube.com/watch?v=xQxf5F f7cQ4 [Accessed: May 24, 2018]

[11] Nakatsugawa I, Suzuki S. Discharge behavior of water activated magnesium-air battery. In: Proceedings of 130th Japan Institute of Light Metals; 2016. p. 16

[12] Nakatsugawa I, Nakano H. Discharge behavior of water-activated magnesium battery. In: Proceedings of 131st Japan Institute of Light Metals; 2016. p. 44

[13] Nakatsugawa I, Nakano H. Discharge behavior of water-activated magnesium battery. Journal of Japan Institute of Light Metals. 2017;**67**:503-510

[14] King AD, Birbilis N, Scully JR. Accurate electrochemical measurement of magnesium corrosion rates; a combined impedance, mass-loss and hydrogen collection study. Electrochimica Acta. 2014;**121**:394-406. DOI: 10.1016/j.electacta.2013.12.124

[15] Curioni M, Scenini F, Monetta T, Bellucci F. Correlation between electrochemical impedance measurements and corrosion rate of magnesium investigated by real-time hydrogen measurement and optical imaging. Electrochimica Acta. 2015;**166**:372-384. DOI: 10.1016/j.electacta.2015.03.050

[16] Feliu S Jr, El Hadad AA, Barranco V, Llorente I, García-Galván FR, Jiménez-Morales A, Galván JC. Native oxide films on AZ31 and AZ61 commercial magnesium alloys—Corrosion behaviour, effect on isothermal oxidation and sol–gel thin film formation. In: Ahmaz Z, editor. New Trends in Alloy Development, Characterization and Application. Rijeka: IntechOpen; 2015. pp. 97-123. DOI: 10.5772/60721

[17] Mekhalfi H, Chelali N, Benhamimid S, Bahloul A. Recycling of manganese dioxide from spent Zn–MnO_2 cells. Russian Journal of Applied Chemistry. 2015;**88**:879-884. DOI: 10.1134/S1070427215050249

[18] Qu D. Application of a.c. impedance technique to the study of the proton diffusion process in the porous MnO_2 electrode. Electrochimica Acta. 2003;**48**:1675-1684. DOI: 10.1016/S0013-4686(03)00146-4

[19] Orazem ME, Tribollet B. Electrochemical Impedance Spectroscopy. New Jersey: Wiley; 2017. pp. 395-418

[20] The Chemical Society of Japan, editors. Handbook of Chemistry. Pure Chemistry. 4th ed. Vol. II-285. Tokyo: Maruzen; 1993

[21] Song G. Corrosion electrochemistry of magnesium and its alloys. In: Song G, editor. Corrosion of Magnesium Alloys. Cambridge: Woodhead Publishing; 2011. p. 12

[22] Glicksmann R. Anodic dissolution of Magnesium alloys in aqueous salt solutions. Journal of the Electrochemical Society. 1959;**106**:83-88. DOI: 10.1149/1.2427299

[23] Frankel GS, Samaniego A, Birbilis N. Evolution of hydrogen at dissolving magnesium surfaces. Corrosion Science. 2013;**70**:104-111. DOI: 10.1016/j.corsci.2013.01.017

[24] Cano ZP, Kish JR, McDermid JR. On the evolution of cathodic activity during corrosion of magnesium alloy AZ31B in a dilute NaCl solution. Journal of the Electrochemical Society. 2016;**163**:C62-C68. DOI: 10.1149/2.0151603jes

[25] Linden D, Reddy TB. Basic concepts. In: Reddy T, editor. Linden's Handbook of Batteries. 4th ed. New York: McGraw-Hill Education; 2011. pp. 1.3-1.17

[26] Shu C, Wang E, Jiang L, Shu SGC. High performance cathode based on carbon fiber felt for magnesium-air fuel cells. International Journal of Hydrogen Energy. 2013;**38**: 5885-5893. DOI: 10.1016/j.ijhydene.2013.02.093

[27] Yang D, Bhattacharjya D, Inamdar S, Park J, Yu J. Phosphorus-doped ordered mesoporous carbons with different lengths as efficient metal-free electrocatalysts for oxygen reduction reaction in alkaline media. Journal of the American Chemical Society. 2012;**134**:16127-16130. DOI: 10.1021/ja306376s

[28] Gorlin Y, Jaramillo TF. A bifunctional nonprecious metal catalyst for oxygen reduction and water oxidation. Journal of the American Chemical Society. 2010;**132**:13612-13614. DOI: 10.1021/ja104587v

[29] Wu Z, Yang S, Sun Y, Parvez K, Feng X, Müllen K. 3D nitrogen-doped graphene aerogel-supported Fe_3O_4 nanoparticles as efficient electrocatalysts for the oxygen reduction reaction. Journal of the American Chemical Society. 2012;**134**:9082-9085. DOI: 10.1021/ja3030565

[30] Milusheva YD, Boukoureshtlieva RI, Hristov SM, Kaisheva AR. Environmentally-clean Mg-air electrochemical power sources. Bulgarian Chemical Communications. 2011;**43**:42-47

[31] Cao D, Wu L, Sun Y, Wang G, Lu Y. Electrochemical behavior of Mg–Li, Mg–Li–Al and Mg–Li–Al–Ce in sodium chloride solution. Journal of Power Sources. 2008;**177**:624-630. DOI: 10.1016/j.jpowsour.2007.11.037

[32] Yuasa M, Huang X, Suzuki K, Mabuchi M, Chino. Discharge properties of Mg-Al-Mn-Ca and Mg-Al-Mn alloys as anode materials for primary magnesium-air batteries. Journal of Power Sources. 2015;**297**:449-456. DOI: 10.1016/j.jpowsour.2015.08.042

[33] Wang N, Wang R, Peng C, Peng B, Feng Y, Hu C. Discharge behaviour of Mg-Al-Pb and Mg-Al-Pb-In alloys as anodes for Mg-air battery. Electrochimica Acta. 2014;**149**:193-205. DOI: 10.1016/j.electacta.2014.10.053

[34] Richey FW, McCloskey BD, Luntz A. Mg anode corrosion in aqueous electrolytes and Implications for Mg-air batteries. Journal of the Electrochemical Society. 2016;**163**:A958-A963. DOI: 10.1149/2.0781606jes

[35] Nwaogu UC, Blawert C, Scharnagl N, Dietzel W, Kainer U. Effects of organic acid pickling on the corrosion resistance of magnesium alloy AZ31 sheet. Corrosion Science. 2010;**52**:2143-2154. DOI: 10.1016/j.corsci.2010.03.002

[36] Yasuda T, Yamazaki T, Honbo E, Takata K, Ishikubo T, Takatsuji N. Development of Magnesium Fuel Cell, Collection of Academic Papers of Toyama New Industry Organization. Toyama: 2013. pp. 26-33

[37] Höche D, Sviatlana L, Zheludkevich ML. Electrolyte additives for magnesium air batteries. European Patent. EP20160187152

[38] Garra P, Leyssens G, Allgaier O, Schönnenbeck C, Tschamber V, Brilhac JF, Tahtouh T, Guézet O, Allano S. Magnesium/air combustion at pilot scale and subsequent PM and NO_x Emissions. Applied Energy. 2017;**189**:578-587. DOI: 10.1016/j.apenergy.2016.12.069

[39] Sherron M. Smart diaper. US patent. US20100241094A1

[40] Bhatti M, Reyzin I, Enzinna D. Leak detection apparatus for liquid circulation cooling system. US patent. US6826948 B1

[41] Tsang M, Armutlulu A, Martinez A, Herrault F, Allen SAB, Allen MG. A MEMS-enabled biodegradable battery for powering transient implantable devices. In: 2014 IEEE 27th International Conference on MEMS; 2014. pp. 358-361. DOI: 10.1109/MEMSYS.2014.6765650

[42] Edupuganti V, Solanki R. Fabrication, characterization, and modeling of a biodegradable battery for transient electronics. Journal of Power Sources. 2016;**336**:447-454. DOI: 10.1016/j.jpowsour.2016.11.004

Chapter 7

Clinical Applications of Magnesium Hydride

Chung-Hsing Chao

Additional information is available at the end of the chapter

http://dx.doi.org/10.5772/intechopen.79507

Abstract

Oxygen sustains the lives of human a unique element. However, oxygen is unwanted and harmful when it is over tension to introduce oxygen-derived free radicals in the cells. Hydrogen and oxygen are both involved in the genesis of life that exists on Earth and metabolism in vivo, so it is not strange to exert a biological effect. Recently, scientists have found that hydrogen is an essential physiological regulatory function with antioxidant, anti-inflammatory, and antiapoptotic protective effects on cells. Using therapeutic hydrogen includes different delivery methods, such as inhalation of hydrogen gas, oral hydrogen water, and injection of hydrogen-saturated saline. In clinical applications, magnesium hydride hydrolysis instead of traditional metallic magnesium is effective in increasing the theoretical hydrogen storage capacity by two times. The hydrogen-water comes across as being an excellent choice to produce from magnesium hydride hydrolysis on-demand because its antioxidant activity cannot store.

Keywords: hydrogen, method, metallic magnesium hydrolysis, magnesium hydride hydrolysis, oral hydrogen-dissolving water, hydrogen gas inhalation, hydrogen-saturated saline injection, reversible osmosis

1. Introduction

In 2007, Oshawa et al. [1] have the first reported that inhalation of only 2% hydrogen gas could selectively reduce oxygen-derived free radicals and improve cerebral ischemia-reperfusion injuries. However, the underlying mechanism is still not precise; the protective effects of the small amount of hydrogen on oxidative stress draw much attention for developing the hydrogen biology. Since that, hydrogen biology becomes a hotspot in the medical research, more than 500 papers related to almost all diseases. Lots of evidence shows molecular hydrogen in cells, and organisms exerting the antioxidant, anti-inflammatory, and antiapoptotic for oxidative stress on some types of infections. Why is this discovery significant? Inhalation of hydrogen below the flammability limit of 4.6%, drinking containing saturated hydrogen water, and injection of containing saturated hydrogen saline could exert therapeutic biological effects for selectively eliminating the toxic free radicals, such as hydroxyl free radicals, superoxide radical in vivo. We can treat the reactive oxygen species-caused oxidative stress disease by administrating reductive drugs, such as vitamins. However, it would eliminate all reactive oxygen species [2]. Cells maintain metabolism through oxidizing food into energy species in vivo. Therefore, cells produce reactive oxygen free radicals to help to absorb nutritious substances, which is a unique phenomenon for biological organisms. But over the tension of free radicals can seriously affect the physiological functions, and even lead to death. Oxidative stress damage likes a peeled apple will turn brown with an injury.

What are free radicals? Free radicals are molecules and atomic groups with "unpaired" electrons. Owing to unpaired electrons, they will snatch free electrons of other molecules in vivo to match with them. After snatching electrons, they will turn molecules of the robbed electrons into newly free radicals. Finally, it will induce the "chain reaction of free radicals" so the unbalance of free radicals will cause the cells crisis. Hydrogen has the lightest weight with one proton and one electron, and the molecular hydrogen a nonradical, nonreactive, nonpolar, highly diffusible neutral gas, which can arbitrarily penetrate the cell membrane into nucleus and mitochondria to neutralize the highly reactive oxygen species in vivo. At a low concentration, hydrogen does not react with the characterized as large in number, low in activity, and beneficial reactive oxygen species, such as hydrogen peroxide, nitric oxide. This phenomenon is called the selectively neutralizing highly reactive free radicals. It means that low concentration hydrogen can be limited to deal with high reactive cytotoxic oxygen species and not to kill all the free radicals. Till now, hydrogen has been shown to have a therapeutic effect on over 170 different human diseases and animal models, including in antioxidant activity [1, 3–9], ischemia-reperfusion injury [2, 10–14], Parkinson's disease [15, 16], cancer [17], retina [12], chondrocytes [7], diabetes disease [18], irradiation-induced lung damage [8], mitochondrial disease [9], and cardiac disease [13, 19].

When hydrogen was focused on its biological effects in multiple diseases by scientists in recent years, the method of hydrogen applications is that we have to resolve it for adequately applying its antioxidant, anti-inflammatory, and antiapoptotic function for oxidative stress [1–19]. This antioxidant effect of molecular hydrogen has been proved. However, molecular hydrogen is not the same as hydrogen gas. Hydrogen water contains both atomic and molecular hydrogen. It is essential continuously to study the medical effects in vivo for hydrogen bioresearch. Producing high-quality hydrogen water method for clinical applications is one of the critical points in the study. Although, many scientists have studied hydrogen-associated techniques including magnesium rod hydrolysis, electrolytic water containing sodium chloride, and electrolyzed water by polymer membrane patents [20–22]. We found that the related methods of producing hydrogen dissolving in water were still limited. This chapter shows a new way of magnesium hydride hydrolysis by reacting with tap water. It is provided to increasing the theoretical hydrogen production amount by more than two times, solving the drawbacks that the rate of metallic magnesium rod hydrolysis is too low, the risk of internal explosion of the hydrogen-oxygen electrolysis process itself, and the cost of pure water electrolysis by polymer membrane method is too high. The magnesium hydride hydrolysis belongs to alkaline water efficiently enhances the saturated hydrogen dissolving in water, and the smallest hydrogen molecule passes nanometer-filter of reversible osmosis transformed into fresh high-content hydrogen water.

2. Materials and methods

At room temperature, the concentration of hydrogen molecule dissolving in tap water is only 8.65×10^{-7} mg/L. At such a low level of hydrogen, there is no therapeutic effect in clinical applications. The saturation-hydrogen of 1.6 milligram/Liter [20–22] in water is more than million times of hydrogen concentration in tap water. This value is considered the hydrogen dissolved in water at room temperature to its maximum state. The amount of hydrogen will reduce if hydrogen escapes from the water surface equaling to the pressure at sea level, which is 101.325 kilopascal (one-atmosphere pressure). Hydrogen, under saturation condition, has been found to be able to exert its protective effects in the biological system [1–19]. Hydrogen is the minimum mass gas molecule with relatively steady reducibility, a valuable chemical property of redox reaction in metabolism. It has been accepted that hydrogen cannot directly interact with substances in the biological body, and that is why, we get hydrogen dissolving in water by oral or injection with saturated hydrogen saline. Since 2007, plenty of trials have been done using hydrogen inhalation medicine in treating ischemia-reperfusion injury at 1 or 2% hydrogen concentration for 35 min. Subsequently, a lot of researchers have shown that a small level of hydrogen has therapeutic effects on a variety of diseases. Even in pure oxygen gas environment, it is not flammable or explosive when the hydrogen concentration is less than 4% or over 95%. **Table 1** shows several delivery methods for hydrogen administration in vivo. In the following section, we introduce the means of producing hydrogen molecules, such as hydrogen from metallic magnesium rod hydrolysis [20], sodium chloride ionic water-electrolyzed hydrogen [21], and pure water through polymer membrane-electrolyzed hydrogen [22]. Later, we will explain and compare them with a specific mention as below.

Administration	Delivery methods	Characteristics
Hydrogen gas	Inhalation gas mixture of $H_2 < 4\%$	Rapid, unsafe, acute oxidative damage
Hydrogen water	Oral intake H_2 water <1.6 mg/L	Low cost, safe, convenient
Hydrogen saline	Intravenous injection by saturated H_2 saline	Accurate dosage
Hydrogen solution	Eye drop, bath immersion	Low cost, safe, convenient
Increased intestinal hydrogen amount	Drug	low cost, convenient

Table 1. Hydrogen delivery methods.

2.1. Magnesium hydrolysis

Metallic magnesium is one of the soluble materials in our body, such as bone material and blood vessel stents. It is safe to use as a medical therapy. However, magnesium reacts violently with steam at high temperatures, such as burning or explosion with water vapor. It is active as a reductant agent for oxidizing water into hydrogen with a standard oxidation potential of +2.356 V. If we throw a small piece of magnesium into cold water, it reacts very slightly. Hydrogen gas is slowly released, because magnesium hydroxide is coated on its surface in the reaction. White-colored magnesium hydroxide mixed with water is called emulsion of magnesium milk which is used to neutralize excess stomach acid in treating constipation disease. Magnesium hydroxide is a base hardly soluble in cold water, while it can react with acid to generate a salt and water in a neutralization reaction. We expressed the chemical response according to the standard reduction potential of magnesium [23].

$$Mg_{(s)} \rightarrow Mg^{+2}{}_{(aq)} + 2e^{-} \; E^{0} = -2.372 \text{ V} \tag{1}$$

$$2H_2O_{(l)} + 2e^{-} \rightarrow H_{2(aq)} + 2OH^{-} \; E^{0} = -0.828 \text{ V} \times 2 \tag{2}$$

Magnesium hydrolysis in redox reaction would be spontaneous ($E > 0$). However, the rate of magnesium hydrolysis is gradually slow down because $Mg(OH)_2$ is highly insoluble in water. Therefore, the total reaction of magnesium hydrolysis is as below.

$$Mg_{(s)} + 2H_2O_{(l)} \rightarrow H_{2(aq)} + Mg\,(OH)_{2(s)} \tag{3}$$

In the solution, two hydroxide ions combine with one magnesium ion to form magnesium hydroxide while leaving insufficient hydroxide ions to conduct the electrons. So, it cannot complete all the hydrolysis reaction despite the escaping hydrogen gas.

Magnesium hydrolysis producing hydrogen-rich water [20] is the most simple and cheap method for oral hydrogen. While drinking hydrogen water, magnesium hydroxide may also take in. There is no potential harm to the human body as a gastric acid inhibitor. If in a day we drink 2 liters of hydrogen water, the intake is 18 mg and will not exceed the limit of 300 mg per day.

2.2. Electrolysis of ionic water

Drinking electrolytic water has used for health care in Japan more than half a century. Commercially available electrolytic water generator [21] is called ion water electrolyzed. The core device of electrolytic water generator includes a pair of anode/cathode electrodes and ionic membrane. The electrolytic reaction has a standard electrode potential of −1.23 V. It means that it at least requires a potential difference of 1.23 V between anode and cathode to split water. The electrochemical reaction includes oxidation at the anode,

$$H_2O_{(l)} \rightarrow 0.5O_{2(aq)} + 2H^{+}{}_{(aq)} + 2e^{-} \; E^{0} = -1.23 \text{ V} \tag{4}$$

reduction at the cathode,

$$2H^{+}_{(aq)} + 2e^{-} \rightarrow H_{2(aq)}\ E^{0} = 0\ V \quad (5)$$

and the overall redox reaction.

$$H_2O_{(l)} \rightarrow 0.5O_{2(aq)} + H_{2(aq)}. \quad (6)$$

Electrolysis of water would be not spontaneous due to $E^0 = -1.23\ V < 0$. Therefore, it needs electrical power to drive electrolyzed water reaction. Electrolysis of water has advantages of simple, low cost, and convenient use. However, if it connects to the tap water, the water may contain part of chlorine or salt residues. The residual chlorine content of electrolysis of water will be 15–20% higher than the original tap water. Apart from chlorine, there is ozone in the water at the anode, and ozone has a pungent odor, a harmful effect on human health. Ozone is a strong oxidant which can also damage the lung causing lung bleeding and death. Therefore, the potential crisis in electrolyzed water generator may originate in improper design or use. For example, many manufacturers to reduce costs will not take the initiative to use gas-tight devices to separate the hydrogen and oxygen further, resulting in hydrogen and oxygen confluence, so that the machine is under explosive risk. Also, the hydrogen gas is generated at the cathode, so the solutions become more and more alkaline. The human body is not suitable for drinking too high-pH-value water. When the alkaline electrolytic water enters the stomach, it will be in a neutralization reaction with acidic gastric acid. Therefore, adjusting the human body's pH value by alkaline electrolytic water is lack of scientific basis.

2.3. Electrolysis of pure water

Recently, the green energy market has sprung up, and a new technology called "polymer proton conducting membrane" of electrolytic production of hydrogen from pure water has emerged. Such electrolyzed pure water is the reverse reaction of "hydrogen fuel cell of power generation," which fuel cell directly transforms the chemical energy of hydrogen and oxygen into electricity. The only by-product is pure water. Japan has spared no effects to promote hydrogen fuel cell technology, TOYOTA, and Honda automotive company have also developed hydrogen fuel cell-powered electric vehicles. It uses pure water and electricity to produce hydrogen and oxygen. Due to high hydrogen production efficiency and high safety factor, the purity of hydrogen can reach over 99.999%. The polymer proton-conducting membrane is manufactured by DuPont Inc. as a solid electrolyte, which does not require dissolved ions in water. Instead, the higher the purity of water, the better can protect the core device of membrane electrode assembly without accompanying other by-products. The electrochemical reaction is as below.

$$H_2O_{(l)} \rightarrow 2H^{+} + 2e^{-} + 0.5O_{2(g)} \text{ at the anode} \quad (7)$$

$$2H^{+} + 2e^{-} \rightarrow H_{2(g)} \text{ at the cathode} \quad (8)$$

Water molecule splits at the anode to form two positively charged hydrogen ions (H^+), two electrons (e^-), and a half oxygen molecule (O_2). The two electrons flow through an external circuit and the two hydrogen ions selectively move across the polymer electrolyte membrane to the cathode. At the cathode, the two hydrogen ions combine with the two electrons from the external circuit to form one hydrogen molecule (H_2). The gas-tight polymer of proton exchange membrane prevents the oxygen at the anode from passing through the layer to the cathode, completely isolating the oxygen from the hydrogen gas, avoiding the internal explosion of the electrolysis process. Electrolytic water maintains at a neutral state of pH equal to 7 during the process of electrolysis. Furthermore, the amount of hydrogen production at a 3 Ampere/cm^2 of current density is five times higher than that of the ion-electrolyzed water. The volume of pure water electrolyzed system can be smaller and lighter. Although the polymer electrolyte membrane has many advantages mentioned above, it is still costly to use. Also, it cannot use tap water to be another problem because metal ion, chlorine, carbon monoxide in the solution will poison the platinum catalyst, and its price cannot compete with the other products.

2.4. Comparison of methods

The method of magnesium hydrolysis of producing hydrogen water was invented by Dr. Hidemitsu Hayashi, who was the heart surgeon being responsible for clinical medicine. He began to study the medical benefits of hydrogen water in 1985 and published a report on hydrogen water treatment 10 years later. The magnesium rod consists of magnesium powders packed in a porous polyethylene resin (see **Figure 1**). The mild chemical reaction between magnesium and water lead to the generation of hydrogen bubbles and the formation of magnesium hydroxide on the surface of magnesium. This simple and effective process produces hydrogen antioxidant solution, which is more alkaline than that of pure water [20].

The ion-electrolyzed water system for producing hydrogen water was Akiyama Hiroyuki's invention [21], and the commercial product of both of acidic water and alkaline water has completed. The electrolytic water system includes an electrolysis generating unit, a reading display, and a control unit. It selectively makes and takes out a plurality of acidic-oxidation water and hydrogen-reduction water with a given pH value which is based on a series of operation modes. The reading unit reads the information stored in the external touch panel carried by the user. The control unit determines the operation mode; the ion electrolysis system is operated to improve the usability by a plurality of people [21].

Wang Dong invented the electrolyzed water system based on the polymer electrolyte membrane [22] as shown in **Figure 2**. The electrolyzed water-generating device provides high-purity hydrogen and oxygen for the fuel cell, including the front end plate, diffusion plate, conductive plate, special plate, flow field plate, membrane electrode, rear special plate, buffer plate, rear end plate, and other units, each unit is pressure-resistant. Water and gas flow field plates and conductive plates made of titanium alloy and other corrosion resistant alloys. The carbon fiber paper is used as the membrane electrode diffusion layer. The membrane electrode is a high-activity electrode prepared by proton exchange

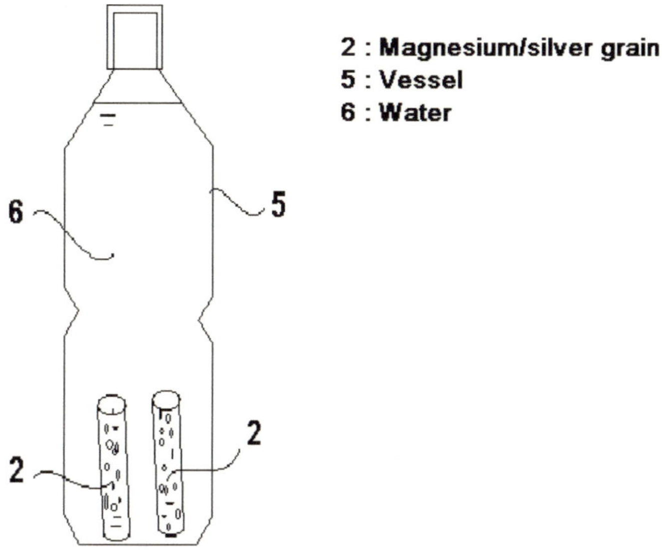

Figure 1. Method of magnesium hydrolysis for producing hydrogen-rich water [20].

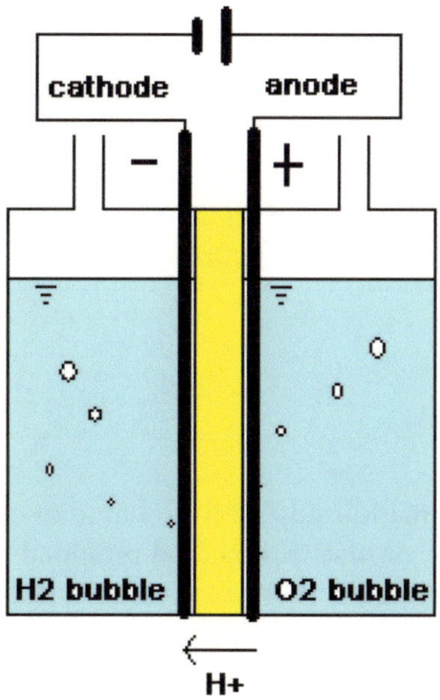

Figure 2. Polymer electrolyte membrane electrolyzer [22].

membrane and composite platinum metal catalyst and carbon black. The working medium of the invention is pure water, has no pollution to the environment, can recycle, and the generated hydrogen water is high in purity and does not need complicated purification and treatment.

Methods	Magnesium hydrolysis [20]	Ion-water electrolysis [21]	Pure-water electrolysis [22]
Theory	Magnesium hydrolysis	Electrolyzed water	Electrolyzed water
Purity	100%	70–98%	100%
Reaction rate	Low	Medium	Medium
Water	Tap water	Ion water	Pure water
By-product	Mg^{+2}, OH^-, $Mg(OH)_2$	Cl_2, HClO, O_2, O_3	None
Safety	High	Explosion, toxicity may be possible	High
Cost	Low	Medium	High

Table 2. Three primary producing hydrogen water methods [20–22].

Table 2 listed the comparison between three primary producing hydrogen-water methods. This seemingly theoretical capacity of hydrogen, purity, reaction rate, etc. indeed relevant for real clinical application scenarios. As in many cases, the water of various sources whether or not is readily on site or therefore needs to transport the necessary amount of pure water or distilled water for use. We attempt to improve the conventional methods of producing hydrogen-water by using magnesium hydride hydrolysis. The double capacity of hydrogen sources of magnesium hydride hydrolysis originates as half from the hydride ions and half of the water protons. Thermodynamically, magnesium hydride on contacting with water to produce hydrogen-water is favorable and the by-product of magnesium hydroxide. However, in practice, not only the reaction kinetics is extremely slow but also the insoluble passivation layer of magnesium hydroxide soon coated on the outer surface to impede water entering. For example, the reaction yield of magnesium hydride hydrolysis at room temperature is below 1% after half an hour.

3. Results and discussion

In this section, we give a systematic study on the method of magnesium hydride hydrolysis for producing hydrogen-water. At first, this method provided to an increasing the theoretical hydrogen amount by two times than that of the technique of magnesium hydrolysis. The hydrogen-water produced from this method belongs to alkaline water. At room temperature, the reaction is prolonged because the low reaction kinetics and the surface are gradually covered with an insoluble layer of hydroxide on the outer to hinder response continuously. However, it can activate by using citric acid as a chelating agent. Like, citric acid can dissolve metal oxides by making them dissoluble. Citric acid belongs to a weak organic acid that has the chemical formula $C_6H_8O_7$. In biochemistry, the citric acid acts as an intermediate in the citric acid cycle occurred in the metabolism of all aerobic organisms. The chemical reaction of the citric acid cycle is the source of two-thirds of the food-derived energy in higher living creatures. Hans Adolf Krebs also received the 1953 Nobel Prize in Physiology Medicine for the discovery. In this study, citric acid acts as both of a catalyst to activate reaction kinetics and a chelating agent to inhibit the magnesium hydroxide formation over the complete course of the reaction, which is the limiting step of magnesium hydride hydrolysis. These effects are advantageous. The by-product, citrate magnesium, also is a vital component of bone in the human body.

The system of producing hydrogen-water, as shown in **Figure 3**, includes a reactor, reverse osmosis unit, valve, pump, control unit, digital display, water input, hydrogen-water output, hydrogen-water for therapeutic use, magnesium hydride block, and tap water input. **Figure 4** shows the porous structure of magnesium hydride of 35 × 35 × 17 in millimeters which was manufactured by direct hydrogenation processes using a low-cost combustion synthesis [24, 25]. The block of magnesium hydride is made by a light gray crystalline powder (99.9% in purity by mass) under high pressure and temperature of 200 bars and 500°C with magnesium iodide catalysts. The porous structure facilitates the water molecules to enter into magnesium hydride inside to generate hydrogen.

The chemical reaction is expressed below.

$$MgH_{2(s)} \rightarrow Mg^{+2}{}_{(aq)} + 2H^{-} \tag{9}$$

The magnesium hydride hydrolysis reaction shows a similar qualitative process. At the beginning of the response, water molecule reacts with magnesium hydride to release magnesium ions and negative hydrogen ions. And then, the negative hydrogen ions continuously split water molecules releasing hydrogen gas and hydroxide ions. Therefore, we can observe a steep increase in pH value.

$$2H^{-} + 2H_2O_{(l)} \rightarrow H_{2(aq)} + 2OH^{-} \tag{10}$$

While the reaction reached the observed equilibrium of pH 11.0, more and more magnesium hydroxide precipitated on the outer surface of the magnesium hydride from the initially supersaturated solutions state. It is in excellent agreement with the observed pH of saturated solutions of magnesium hydroxide in the literature [23, 25].

$$Mg^{+2}{}_{(aq)} + 2OH^{-} \rightarrow Mg\,(OH)_{2(s)} \tag{11}$$

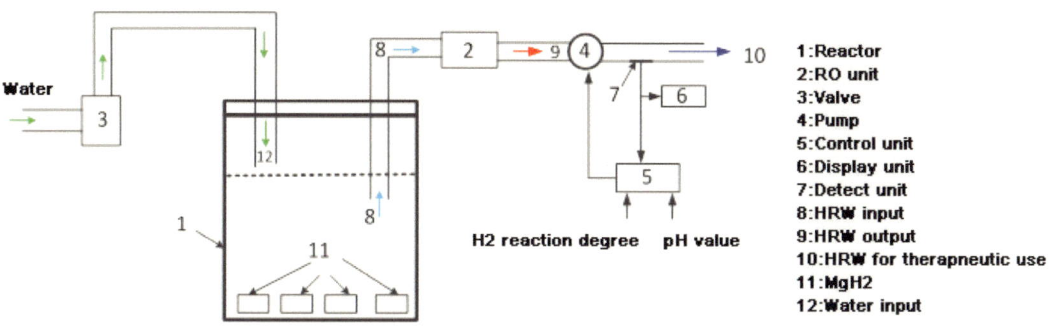

Figure 3. The method of magnesium hydride hydrolysis producing hydrogen-water.

Figure 4. Magnesium hydride (the external size of 35 × 35 × 17 millimeter, Bio-Coke Ltd. Japan).

That is to say, if we carefully control the pH in the hydrolysis reaction less than 11 with magnesium ions concentration less than 0.007 mol per liter, the supersaturated solution state in the solution will not occur. The insoluble magnesium hydroxide layer will not form on the surface to impede hydrolysis reaction. Magnesium hydride is split extremely slow to the hydrogen-saturated state (0.8 mM). The pH-dependence of the hydrolysis behavior of magnesium hydride seems reasonable. Furthermore, we repeatedly inject a small amount of citric acid solution before supersaturation state. The citric acid as a chelating agent can effectively reduce the concentration of magnesium ions and hydroxide ions to generate the soluble magnesium citrate.

$$3MgH_{2(s)} + 2\,C_3H_4(OH)(COOH)_{3(aq)} \rightarrow 6H_{2(aq)} + Mg_3\left[C_3H_4(OH)(COO)_3\right]_{2(aq)} \tag{12}$$

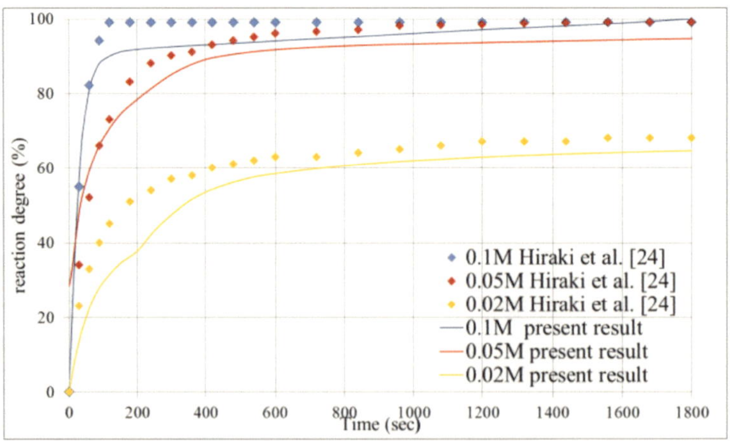

Figure 5. Reaction degree measured in the present result, which also compared with the data of [26].

The reaction percentage was calculated by:

Reaction degree in % = H_2/H_2 in theoretical value in Eq. (12).(13)

The above behavior has also confirmed in our experiments with a stoichiometric amount of citric acid. For applications, the system injects a small amount of citric acid solution by a sequence of multiple separate times of various steps. To overcome the insoluble passivation layer, the mole ratio of citric acid to magnesium hydride is 2:3. About five times the mass of citric acid compared to magnesium hydride is necessary. **Figure 5** shows the measured reaction degree of magnesium hydride hydrolysis for 0.02, 0.05, and 0.1 moles of citric acid per one-liter-pure water. This data also compared with the literature of Hiraki et al. [26]. The time of dissolving molecular hydrogen as shown in the figure was sufficient for continuous use of 30 min for clinical applications. Finally, citric acid can completely prevent the insoluble passivation layers of magnesium hydroxide and thus significantly increase the reaction kinetics and yield.

4. Conclusions

The magnesium hydride hydrolysis is a very promising method to generate saturated hydrogen water since it is cheap, nontoxic, with a very long shelf life to give a two-times storage capacity of hydrogen than that of metallic magnesium. However, pure magnesium hydride hydrolysis reacts extremely slow and incompletely with water. At room temperature, the yield of hydrogen is less than 1% after half an hour. The addition of a small amount of citric acid solution highly increases the reaction kinetics and hydrogen yield. The kinetics and yield strongly depend on the pH in the solution, which can be adjusted by acids. This technology can serve as a superior replacement for magnesium hydrolysis, avoiding the danger of magnesium burning in oxygen. It is also a significant weight reduction in clinical applications while comparing with other methods of generating hydrogen-water. However, a small amount of molecular hydrogen on oxidative stress by efficiently neutralizing reactive oxygen species only explains part of the phenomena. Now, it is challenged by increasingly accumulated evidence. The future needs to prove the clinical effectiveness, and the evidence in the hydrogen molecular mechanism, medical, and biological effects.

Author details

Chung-Hsing Chao

Address all correspondence to: dd500929@gmail.com

Ta Hwa University of Science and Technology, Hsinchu County, Taiwan R.O.C

References

[1] Ohsawa I, Ishikawa M, Takahashi K, Watanabe M, Nishimaki K, Yamagata K, Katsura K, Katayama Y, Asoh S, Ohta S. Hydrogen acts as a therapeutic antioxidant by selectively reducing cytotoxic oxygen radicals. Nature Medicine. 2007;**13**(6):688-694. DOI: 10.1038/nm1577

[2] Sato Y, Kajiyama S, Amano A, Kondo Y, Sasaki T, Handa S, Takahashi R, Fukui M, Hasegawa G, Nakamura N, Fujinawa H, Mori T, Ohta M, Obayashi H, Maruyama N, Ishigami A. Hydrogen-rich pure water prevents superoxide formation in brain slices of vitamin C-depleted SMP30/GNL knockout mice. Biochemical and Biophysical Research Communications. 2008;**375**(3):346-350. DOI: 10.1016/j.bbrc.2008.08.020

[3] Fukuda KI, Asoh S, Ishikawa M, Yamamoto Y, Ohsawa I, Ohta S. Inhalation of hydrogen gas suppresses hepatic injury caused by ischemia/reperfusion through reducing oxidative stress. Biochemical and Biophysical Research Communications. 2007;**361**(3):670-674. DOI: 10.1016/j.bbrc.2007.07.088

[4] Nagata K, Nakashima-Kamimura N, Mikami M, Ohsawa I, Ohta S. Consumption of molecular hydrogen prevents the stress-induced impairments in hippocampus-dependent learning tasks during chronic physical restraint in mice. Neuropsychopharmacology. 2009;**34**(2):501-508. DOI: 10.1038/npp.2008.95

[5] Ohsawa I, Kiyomi N, Kumi Y, Masahiro I, Shigeo O. Consumption of hydrogen water prevents atherosclerosis in apolipoprotein E knockout mice. Biochemical and Biophysical Research Communications. 2008;**377**(4):1195-1198. DOI: 10.1016/j.bbrc.2008.10.156

[6] Ohta S, Nakao A, Ohno K. The 2011 medical molecular hydrogen symposium: An inaugural symposium of the journal medical gas research. Medical Gas Research. 2011;**1**(1):1-10. DOI: 10.1186/2045-9912-1-10

[7] Hanaoka T, Kamimura N, Yokota T, Takai S, Ohta S. Molecular hydrogen protects chondrocytes from oxidative stress and indirectly alters gene expressions through reducing peroxynitrite derived from nitric oxide. Medical Gas Research. 2011;**1**(1):1-18. DOI: 10.1186/2045-9912-1-18

[8] Terasaki Y, Ohsawa I, Terasaki M, Takahashi M, Kunugi S, Dedong K, Urushiyama H, Amenomori S, Kaneko-Togashi M, Kuwahara N, Ishikawa A, Kamimura N, Ohta S, Fukuda Y. Hydrogen therapy attenuates irradiation-induced lung damage by reducing oxidative stress. American Journal of Physiology. Lung Cellular and Molecular Physiology. 2011;**301**(4):L415-L426. DOI: 10.1152/ajplung.00008.2011

[9] Ohta S. Molecular hydrogen is a novel antioxidant to efficiently reduce oxidative stress with potential for the improvement of mitochondrial diseases. Biochimica et Biophysica Acta. 2012;**1820**(5):586-594. DOI: 10.1016/j.bbagen.2011.05.006

[10] Hayashida K, Sano M, Ohsawa I, Shinmura K, Tamaki K, Kimura K, Endo J, Katayama T, Kawamura A, Kohsaka S, Makino S, Ohta S, Ogawa S, Fukuda K. Inhalation of

hydrogen gas reduces infarct size in the rat model of myocardial ischemia–reperfusion injury. Biochemical and Biophysical Research Communications. 2008;**373**(1):30-35. DOI: 10.1016/j.bbrc.2008.05.165

[11] Cai J, Kang Z, Liu WW, Luo X, Qiang S, Zhang JH, Ohta S, Sun X, Tao H, Li R. Hydrogen therapy reduces apoptosis in neonatal hypoxia–ischemia rat model. Neuroscience Letters. 2008;**441**(2):167-172. DOI: 10.1016/j.neulet.2008.05.077

[12] Oharazawa H, Igarashi T, Yokota T, Fujii H, Suzuki H, Machide M, Takahashi H, Ohta S, Ohsawa I. Protection of the retina by rapid diffusion of hydrogen: Administration of hydrogen-loaded eye drops in retinal ischemia–reperfusion injury. Investigative Ophthalmology & Visual Science. 2010;**51**(1):487-492. DOI: 10.1167/iovs.09-4089

[13] Hayashida K, Sano M, Kamimura N, Yokota T, Suzuki M, Maekawa Y, Kawamura A, Abe T, Ohta S, Fukuda K, Hori S. H2 gas improves functional outcome after cardiac arrest to an extent comparable to therapeutic hypothermia in a rat model. Journal of the American Heart Association. 2012;**1**(5):1-13. DOI: 10.1161/JAHA.112.003459

[14] Cui J, Chen X, Zhai X, Shi D, Zhang R, Zhi X, Li X, Gu Z, Weng W, Zhang J, Wang L, Sun X. Inhalation of water electrolysis-derived hydrogen ameliorates cerebral ischemia-reperfusion injury in rats-a possible new hydrogen resource for clinical use. Neuroscience. 2016;**335**:232-241. DOI: 10.1016/j.neurosscience.2016.08.021

[15] Fu Y, Ito M, Fujita Y, Ito M, Ichihara M, Masuda A, Suzuki Y, Maesaea S, Kajita Y, Hirayama M, Ohsawa I, Ohta S, Ohno K. Molecular hydrogen is protective against 6-hydroxydopamine-induced nigrostriatal degeneration in a rat model of Parkinson's disease. Neuroscience Letters. 2009;**453**(2):81-85. DOI: 10.1016/j.neulet.2009.02.016

[16] Yoritaka A, Takanashi M, Hirayama M, Nakahara T, Ohta S, Hattori N. Pilot study of H2 therapy in Parkinson's disease: A randomized double blind placebo-controlled trial. Movement Disorders. 2013;**28**(6):836-839. DOI: 10.1002/mds.25375

[17] Nakashima-Kaminura N, Mori T, Ohsawa I, Asoh S, Ohta S. Molecular hydrogen alleviates nephrotoxicity induced by an anti-cancer drug cisplatin without compromising anti-tumor activity in mice. Cancer Chemotherapy and Pharmacology. 2009;**64**(4):753-761. DOI: 10.1007/s00280-008-0924-2

[18] Kamimura N, Nishimaki K, Ohsawa I, Ohta S. Molecular hydrogen improves obesity and diabetes by inducing hepatic FGF21 and stimulating energy metabolism in db/db mice. Obesity. 2011;**19**(7):1396-1403. DOI: 10.1038/oby.2011.6

[19] Gao Y, Gui Q, Jin L, Yu P, Wu L, Cao L, Wang Q, Duan M. Hydrogen-rich saline attenuates hippocampus endoplasmic reticulum stress after cardiac arrest in rats. Neuroscience Letters. 2017;**640**:29-36. DOI: 10.1016/j.neulet.2017.01.020

[20] Hayashi H. Method of producing hydrogen rich water and hydrogen water generation. US Patent No. 7,189,330 B2. 13 March 2007

[21] Akiyama Hiroyuki (秋山博之). Electrolyzed Water System (電解水生成系統), Japan Patent (日本特許番號) NO.6232037, 15 November 2017

[22] Wang DZ et al. Water electrolysis device with proton exchange membrane. CN1966777A. 17 November 2005

[23] Mueller WD, Homberger H. The influence of MgH2 on the assessment of electrochemical data to predict the degradation rate of Mg and Mg alloys. International Journal of Molecular Sciences. 2014;**15**(7):11456-11472. DOI: 10.3390/ijms150711456

[24] Macdonald D. OSTI.GOV/Report Number DOE/GO--15054: Electrochemical Hydrogen Storage Systems. Publication Date: 9 August 2010. DOI: 10.2172/984730

[25] Uesugi H, Sugiyama T, Nii H, Ito T, Nakatsugawa I. Industrial production of MgH2 and its application. Journal of Alloys and Compounds. 2011;**509**:S650-S653. DOI: 10.1016/j.jallcom.2010.11.047

[26] Hiraki T, Hiroi S, Akashi T, Okinaka N, Akiyama T. Chemical equilibrium analysis for hydrolysis of magnesium hydride to generate hydrogen. International Journal of Hydrogen Energy. 2012;**37**(17):12114-12119. DOI: 10.1016/j.ijhydene.2012.06.012